SANTA ANA PUBLIC LIBRARY

D0954776

 # About Island Press

Since 1984, the nonprofit organization Island Press has been stimulating, shaping, and communicating ideas that are essential for solving environmental problems worldwide. With more than 1,000 titles in print and some 30 new releases each year, we are the nation's leading publisher on environmental issues. We identify innovative thinkers and emerging trends in the environmental field. We work with world-renowned experts and authors to develop cross-disciplinary solutions to environmental challenges.

Island Press designs and executes educational campaigns in conjunction with our authors to communicate their critical messages in print, in person, and online using the latest technologies, innovative programs, and the media. Our goal is to reach targeted audiences—scientists, policymakers, environmental advocates, urban planners, the media, and concerned citizens—with information that can be used to create the framework for long-term ecological health and human well-being.

Island Press gratefully acknowledges major support of our work by The Agua Fund, The Andrew W. Mellon Foundation, The Bobolink Foundation, The Curtis and Edith Munson Foundation, Forrest C. and Frances H. Lattner Foundation, The JPB Foundation, The Kresge Foundation, The Oram Foundation, Inc., The Overbrook Foundation, The S.D. Bechtel, Jr. Foundation, The Summit Charitable Foundation, Inc., and many other generous supporters.

The opinions expressed in this book are those of the author(s) and do not necessarily reflect the views of our supporters.

Three Revolutions

Three Revolutions

Steering Automated, Shared, and Electric Vehicles to a Better Future

Daniel Sperling (signature)

Daniel Sperling

with contributions by Anne Brown, Robin Chase,

Michael J. Dunne, Susan Pike, Steven E. Polzin, Susan Shaheen,

Brian D. Taylor, Levi Tillemann, and Ellen van der Meer

⬤ **ISLAND**PRESS

Washington | Covelo | London

Library of Congress Control Number: 2017951060

All Island Press books are printed on environmentally responsible materials.

Manufactured in the United States of America
10 9 8 7 6 5 4 3 2 1

Keywords: Driverless car, electrification, energy policy, electric vehicle (EV) batteries, fleet logic, Google car, GHG emissions, Lyft, microtransit, mobility, paratransit, parking, pooling, ridehailing, self-driving car, Tesla, traffic congestion, transit, transportation policy, Uber, vehicle technology, zero-emission vehicle (ZEV)

Contents

Preface

Creating this book has been a journey. Like any book, it emerged from historical observations and personal experiences. My first jobs, as an urban planner in the Peace Corps in Tegucigalpa, Honduras, and at the US Environmental Protection Agency (EPA) for nearly two years, opened my eyes to the social and environmental underbelly of economic growth as well as the opportunities and failures of policy and government.

The "synfuel" fiascoes of the late 1970s and early 1980s, which I observed as an impressionable graduate student, were a lesson in how even the most enlightened industry and government leaders can get it wrong. Then came the so-called intelligent vehicle and highway systems (IVHS) technologies of the 1990s, which emerged from a narrow community of government traffic managers and companies interested in vehicle gadgets. So much promise, and so little impact.

This long series of largely negative and discouraging experiences came to an end in the late 1990s, when Susan Shaheen, then my PhD student, and I helped broaden the IVHS community to embrace carsharing. With a $500,000 gift from Honda Motor Company, we launched the Center for New Mobility Studies at UC Davis. Susan carried the flag as cofounder of the Transportation Sustainability Center at the University of California, Berkeley, but it was a lonely time for her—and for me. We were fifteen years ahead of our time. I moved on.

Meanwhile, I had founded the Institute of Transportation Studies at UC Davis in 1991 with the goal of advancing research that would steer transportation investments and innovations toward environmental sustainability. The institute has exceeded even my most inflated expectations. We were chosen to hostz the National Center for Sustainable Transportation in 2013 and then again in 2016 and have built the research expertise to undergird an array of important government policies, from zero-emission vehicle (ZEV) mandates and low-carbon fuel standards to policies regulating shared mobility services. The hundreds of faculty members and students affiliated with the institute over the years have provided the expertise and inspiration for this book.

I am now more optimistic than ever before, despite the trauma around the world and the political and policy breakdowns here in the United States. Why? My professional dreams and aspirations are now being realized. For the first time in half a century, real transformative innovations are coming to our world of passenger transportation—with the promise of huge energy, environmental, and social benefits. Those decades in the wilderness of stagnation are now being swept away. The narrow vision of IVHS, which later evolved into a slightly broader intelligent transportation system (ITS) vision, is now in full bloom as the three transportation "revolutions" of vehicle electrification, shared mobility, and vehicle automation.

Policy is critical in each. It is tugging vehicle electrification into the marketplace, and it is guiding the other two revolutions toward the public interest. Not since the advent of interstate highways in the United States, high-speed rails in Japan and Europe, and containerized movement of goods have we seen such transformative opportunities. What an exciting and important time!

This book gelled in early 2016, when Anthony Eggert of the Climateworks Foundation and Patty Monahan of the Energy Foundation approached me about running a policy conference on what I had been calling the three mobility revolutions—first in my January 2015 Thomas Deen Distinguished Lectureship for the Transportation Research Board in Washington, DC, and then more explicitly in a September 2015 keynote presentation to the Society of Automotive Engineers in Chicago. I added this book to the initiative and recruited leading experts to coauthor the chapters.

For purposes of this book, we are defining *revolution* as "a fundamental change in the way of thinking about or visualizing something" and "a changeover in use or preference, especially in technology," as *Webster's* does. The chapters on electrification, shared use, and automation are arranged in order of the revolutionary potential of each of these innovations. Subsequent chapters explore the implications for public transit, equity, and the auto industry and how China will (or might) benefit from the three revolutions.

Electrification is discussed first because electric vehicles (EVs) are artifacts that replace another artifact—internal combustion engine vehicles—with only modest ripple effects through the economy and society. Indeed, the electrification of our economy goes back to the late 1800s. Electrification of vehicles probably doesn't even merit the label "disruptive innovation," using the language of Clayton Christensen. Yes, EVs affect the business model of automakers, but their manufacture and sale are unfolding slowly over time, and the core business of automakers is not threatened—for most companies. Car companies are simply switching to a new power plant. EVs might eventually bring about "a changeover in use or preference," but those changes in use won't happen suddenly or completely—and in any case, those uses will be shaped more by the other two innovations. How the other two innovations will unfold is less certain, but both have the potential to be far more transformative—even revolutionary.

My epiphany, midway through writing this book, was that the path to sustainable transportation begins with pooling: filling our cars, buses, and trains with more passengers. Pooling has a greater potential than

EVs to transform transportation. Indeed, I've come to believe that it is the single most important strategy and innovation going forward for all passenger transportation. It builds not on conventional carpooling, which has been a failure, but on what I call the glorified taxi services of Lyft and Uber (and others in the United States and elsewhere).

UberX and traditional Lyft services are a necessary foundation for the embrace of new services that carry multiple riders—think Lyft Line and UberPool, as well as microtransit services. In this new world of pooling, automakers will refashion themselves as mobility service providers. Individual car ownership will start to dwindle. But change will be slow, and the benefits will be modest. The key to massive transformation, with the potential for huge benefits—or huge degradation—is the third innovation, vehicle automation.

Driverless cars will greatly leverage all the benefits of pooling. When cars are fully driverless—as they surely will be someday—the cost of travel in time and money will sharply diminish.

The overall benefits to society of the three revolutions will be massive: trillions of dollars in cost savings globally, sharply reduced greenhouse gas emissions, and more mobility for more people, including those who are now mobility disadvantaged. I'm pretty sure we will get there, eventually.

But then again, I tend to see a glass half full.

A less cheerful scenario—the glass half empty view—is also plausible. In this case, cars remain individually owned, vehicle electrification proceeds slowly, people resist pooling, and government policy makers are passive. In this automation-dominant scenario, with minimal pooling and lagging electrification, vehicle use and greenhouse gas emissions soar, the gap between haves and have-nots widens, and the social and economic fabric of society unravels.

Many factors will influence the unfolding of these transformations— these revolutions—including the willingness of travelers to share rides and eschew car ownership; continuing reductions in battery, fuel cell, and automation costs; and the adaptiveness of companies. But one of the most important factors is policy.

Policy plays a central role in accelerating these innovations. And policy is *the* most important factor in making sure the innovations serve the public interest—with a transportation system built on shared, electric, automated vehicles. I take the hopeful view that we will bring science and the lessons of the past to bear and will use policy to guide automation toward the public interest and a better future.

This book is a step in that direction. We've woven together the thoughts and writings of a brilliant, expert group of contributors to advise us of potential pitfalls and detours and to steer us toward the most positive scenario. The book is aimed at the wide variety of professionals and educated members of the public who are curious about the future of transportation, our cities, and the environment. These include federal, state, and local government officials responsible for formulating policy in the face of these three revolutions, automotive professionals trying to understand the implications for their industry, professors and students around the world trying to stay abreast of these many changes, and all of us frustrated by bumper-to-bumper traffic, apprehensive about the social well-being of our communities, and concerned about the future of our planet.

As for me, I'm devoting the rest of my professional career to bringing science to policy and steering these revolutions toward the public interest and a better future. I'm working with others to create a "Three Revolutions" policy platform at UC Davis to network all those committed to the cause.

Daniel Sperling, Davis, California

Acknowledgments

My deepest appreciation and admiration go to Lorraine Anderson, who was my supereditor. She managed the collaborations with my coauthors from the very beginning, reined in my digressions, helped reorganize material, and tightened up my writing. Without her, the book would have stretched twice as long in time and words.

My coauthors are the other indispensable and valued partners in this exercise. We worked as a team, testing new ideas and weaving those ideas and themes through the entire book. The whole experience was inspiring and enjoyable.

The graphics for the book were expertly drawn and prepared by Emmanuel Franco and Kelly Chang, undergraduate students with the Institute of Transportation Studies at UC Davis, under the watchful eye of Steve Kulieke, the communications director at the institute.

Aditi Meshram, one of my outstanding graduate students, calculated the costs of automated, pooled, and electric vehicles for this book, and Jinpeng Gao, another one of my outstanding graduate students, provided significant help with references and data, especially for chapter 8.

I am grateful to Anthony Eggert and Patty Monahan for asking me to lead a conference on the three revolutions (November 2016), which became the launchpad for this book, and for their faith and trust that this project merited their support.

I also appreciate Heather Boyer, my editor at Island Press, who saw the value of this work from the first moment and remained a strong advocate for the book throughout the process.

Many others helped as reviewers and alerted me to new references for the rapidly evolving topics in this book. Austin Brown, who joined me at UC Davis as the book was nearing completion, was our go-to expert on policy and along with Mollie D'Agostini will be leading follow-up efforts to implement the many policies suggested in this book. Both provided many valuable suggestions (and corrections) and were careful reviewers of several chapters. Other chapter reviewers included Alberto Ayala, Joshua Cunningham, Gil Tal, Tom Turrentine, and Yunshi Wang. My colleague Lew Fulton helped frame this book by conducting a full-blown international analysis of the impacts of the three revolutions. I owe an intellectual debt to Alain Kornhauser for his stream of insightful blogs about automated vehicles. Ongoing insights (and criticisms) were provided by Larry Burns, Emily Castor, Neil Pedersen, Andrew Salzberg, John Viera, and my entire spring 2017 graduate class of thirty, who wrote a series of papers poking holes in many of the ideas in the book but also proposing imaginative solutions.

The most important person through the whole process (well, other than Lorraine) was my partner and wife, Sandy Berg. She was my champion, sounding board, and occasional skeptic.

CHAPTER 1

Will the Transportation Revolutions Improve Our Lives—or Make Them Worse?

Daniel Sperling, Susan Pike, and Robin Chase

We must steer oncoming innovations toward the public interest—toward shared, electric, automated vehicles. If we don't, we risk creating a nightmare.

WE LOVE OUR CARS. OR AT LEAST we love the freedom, flexibility, convenience, and comfort they offer. That love affair has been clear and unchallenged since the advent of the Model T a century ago. No longer. Now the privately owned, human-driven, gasoline-powered automobile is being attacked from many directions, with change threatening to upend travel and transportation as we know it. The businesses of car making and transit supply—never mind taxis, road building, and highway funding—are about to be disrupted. And with this disruption will come a transformation of our lifestyles. The signs are all around us.

Maybe you use Zipcar, Lyft, or Uber or know someone who does. You've probably seen a few electric vehicles (EVs) on the streets, mostly

Nissan Leafs, Chevy Volts and Bolts, Teslas, and occasionally others. And you've undoubtedly heard and read stories about self-driving cars coming soon and changing everything. But how fast are the three revolutions in electric, shared, and automated vehicles happening, and will they converge? Will EVs become more affordable and serve the needs of most drivers? Will many of us really be willing to discard our cars and share rides and vehicles with others? Will we trust robots to drive our cars?

We're at a fork in the road.

Over the past half century, transportation has barely changed. Yes, cars are safer and more reliable and more comfortable, but they still travel at the same speed, still have the same carrying capacity, and still guzzle gasoline with an internal combustion engine. Public transit hasn't changed much either, though modern urban rail services have appeared in some cities since the 1970s. Likewise, roads are essentially unchanged, still made with asphalt and concrete and still funded mostly by gasoline and diesel taxes. We have a system in which our personal vehicles serve all purposes, and all roads serve all vehicles (except bicycles). It is incredibly expensive, inefficient, and resource intensive.

But it's even worse than that. Most cars usually carry only one person and, most wasteful of all, sit unused about 95 percent of the time.[1] As wasteful and inefficient as they are, cars have largely vanquished public transit in most places. Buses and rail transit now account for only 1 percent of passenger miles in the United States.[2] Those who can't drive because they're too young, too poor, or too physically diminished are dependent on others for access to basic goods and services in all but a few dense cities.

Starting in Los Angeles, the United States built this incredibly expensive car monoculture, and it is being imitated around the world. Cars provide unequaled freedom and flexibility for many but at a very high cost. Owners of new cars in the United States spend on average about

$8,500 per vehicle per year, accounting for 17 percent of their household budgets.[3] On top of that is the cost to society of overbuilt roads, deaths and injuries, air pollution, carbon emissions, oil wars, and unhealthy lifestyles. The statistics are mind-numbing. For the United States alone, consider that nearly 40,000 people were killed and 4.6 million seriously injured in 2016 in car, motorcycle, and truck accidents.[4] Nearly ten million barrels of oil are burned every day in the United States by our vehicles.[5] Transportation accounts for a greater proportion of greenhouse gases than any other sector.[6] Farther afield, in Singapore, 12 percent of the island nation's scarce land is devoted to car infrastructure.[7] In Delhi, 4.4 million children have irreversible lung damage because of poor air quality, mostly due to motor vehicles.[8] We have created an unsustainable and highly inequitable transportation system.

But change is afoot, finally. For the first time since the advent of the Model T one hundred years ago, we have new options. The information technology revolution, which transformed how we communicate, do research, buy books, listen to music, and find a date, has finally come to transportation. We now have the potential to transform how we get around—to create a dream transportation system of shared, electric, automated vehicles that provides access for everyone and eliminates traffic congestion at far less cost than our current system. Or not. It could go awry. It could turn out to be a nightmare.

Let's take a minute to imagine two different scenarios set in the year 2040.

Transportation 2040: The Dream

In one vision of the future, the government has managed to steer the three revolutions toward the common good with forward-thinking strategies and policies. Citizens have the freedom to choose from many clean transportation options. They can spend their time with family and friends rather than in traffic thanks to pooled automated cars. They breathe cleaner air, worry less about greenhouse gas emissions, and trust that transportation is safer, more efficient, and more accessible than ever before. The search for parking is an inconvenience of the past. Worries about Grandma being homebound have evaporated. No longer must parents devote hours to ferrying their kids everywhere. Transportation innovations have made it easy for people to meet all their transportation needs conveniently and at a reasonable cost.

On a typical day in this optimistic scenario, Patricia Mathews and Roberto Ruiz eat breakfast at home with their two children before Pat is picked up by an electric automated vehicle (AV) owned by a mobility company. The AV is dispatched from a mobility hub, where trains come and go, bikes are available, and AVs pick up and drop off passengers.

Like most homes in the neighborhood, the Mathews-Ruiz home has a small pickup area and vegetable garden in front, replacing what had been a large driveway. The garage has been converted to a guest room. Parks and public gardens are connected in a greenbelt that runs behind the homes. Children scamper around without parents worrying about traffic.

As Pat approaches the dispatched AV, it recognizes her and opens a door. Her unique scan authorizes a secure payment mediated through blockchain from her family's mobility subscription account, which also pays for transit, bikeshare, and other transportation services. For a small monthly fee, plus a per-mile charge, the family gains access to a variety of shared vehicles and services, including AVs, electric scooters, and inter-city trains. Discounts are also available for special services like air travel. The account isn't connected to a traceable bank account, and travel data are erased every two months.

As Pat settles in for the short commute, the AV is notified to pick up one more passenger along the way, a neighbor Pat knows. On the way into the city, they chat about the upcoming neighborhood block party. The AV picks up another passenger and heads to the city center, where it is routed onto a broad boulevard with two lanes for auto travel, a reserved lane in the middle for trucks and buses, and bike lanes on each side, flanked by wide pedestrian walkways.

The rest of the Mathews-Ruiz family heads out on shared-use bicycles to the children's school (with AVs available as backup on rainy days). Roberto continues on to the fitness center where he works. It takes him about twenty minutes. At lunchtime, Roberto will hop on a shared electric bike to meet his mother for lunch on the other side of town. She lives in a little neighborhood with a dense mix of shops and residences. Street parking was removed years ago and replaced with a sprinkling of passenger-loading and goods-delivery spaces, extensive bike pathways, wide sidewalks, outdoor seating, and pocket parks.

After lunch, Roberto helps his mother arrange a ride to a nearby medical center in an AV specially designed for physically limited passengers. On her way home, she will be dropped off to visit one of her friends. Her retirement income easily covers her mobility subscription and gives her and other low-income citizens many options for travel. For those with less income, subscription subsidies are available. The subsidies go further if the travelers use AVs during off-peak hours, when many AVs are being parked and recharged. AV dispatching is optimized to match shifts in demand, and travel is priced accordingly.

Back at work, Pat calls Roberto to make plans for the evening. The kids will return from school with a bicycle group and meet their babysitter. Roberto and Pat will hail an AV to go to dinner at one of the pop-up restaurants in the neighborhood park—knowing that if they drink too much, they won't need to worry about driving themselves home.

Transportation 2040: The Nightmare
Now imagine the very different future that could come about if our community is unprepared for the three revolutions. Instead of adopting policies and incentives to encourage pooling of rides, the city allows the

private desires of individuals and the competitive instincts of automotive companies to prevail. Traffic congestion gets worse as people who can afford AVs indulge themselves and send their cars out empty on errands. Most AVs are not electrified, and greenhouse gas emissions increase as people travel more. Time to spend with children and engage in community service becomes scarce. Transit services diminish as rich commuters abandon buses and rail and withdraw their support for transit. Those without driver's licenses and cars continue to be marginalized as the divide between mobility haves and have-nots becomes a chasm. Meanwhile, suburbs sprawl as people seek affordable homes farther and farther out, opting for long commutes and cheap mortgages over proximity and more expensive real estate.

In this future, the middle-class Mathews-Ruiz family owns their own AV, which they've named Hal after the all-powerful computer in the movie *2001: A Space Odyssey*. They live in the outer reaches of suburbia in a modest home. With vehicles still personal property (a residue of the twentieth century) and public transportation minimal, the family is trapped into spending nearly half their income on their beloved AV. They pay not only the high cost of the vehicle but also substantial expenses for remote parking (when not at home), required software updates, safety checks for software and hardware, and access to special AV lanes.

Their eighty-mile commute to jobs in the city center consumes about an hour and a half each way. When they pay to use a special AV lane, Hal can cruise at eighty miles per hour, but the trip is slowed down by Hal's having to drive on mixed-use lanes to and from the high-speed freeway. And on some days, they opt for the mixed-use lanes on their way to and from work to reduce their toll costs, increasing their commute time to two hours each way.

The family buys an AV lane pass that allows them three thousand miles per month, after which they pay $0.40 per mile. With their eighty-mile commute, they use up the entire allocation each month just getting to and from work, forcing them to pay the higher off-subscription rates for the rest of the miles they log.

On days when one of them works late, they send Hal back empty to fetch that other person, or whoever finishes work earlier bides his or her time meandering in Hal along the streets in the crowded, congested city center. But the car is comfortable and the rider can use the time productively (even napping!).

Hal accrues still more miles traveling to pick up the children in the afternoon and running errands during the day. These errands might include a trip to pick up packages at a warehouse, where Hal is recognized with scanning technology and robots load and unload boxes. With all this travel, plus weekend recreation, Hal's mileage generally exceeds five thousand miles per month.

The Fork in the Road

Will either of these futures materialize? Will the three revolutions usher in more vehicle use, increased urban sprawl, more marginalization of mobility have-nots, more expensive transportation, and higher greenhouse gas emissions? Or will they lead to reduced congestion and environmental impacts, safer communities, and easier and cheaper access for all (see figure 1.1)? The answer is unknown and will likely vary greatly across regions and countries. The three revolutions will unfold at different speeds in different places, creating waves of unintended—or at least unanticipated—consequences. We do have a say, but only if we wake up now to the speed and scope of change and how the coming revolutions will impact mobility and cities. Decisions made now about infrastructure and vehicle technologies will strongly influence the path and speed of change (a concept known as path dependence).

Two of these revolutions—vehicle electrification and automation—are inevitable. The third, pooling, is less certain but in many ways most critical, especially as AVs come into being. All three have the potential

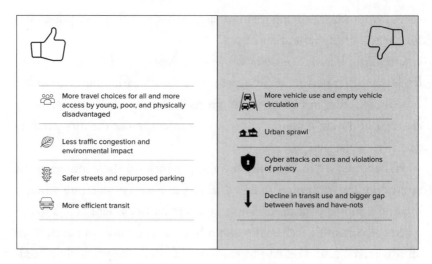

Figure 1.1. The fork in the road: Will transportation in 2040 be a dream or a nightmare?

to offer large benefits. EVs, including those powered by hydrogen, will decrease the use of fossil fuels and lower greenhouse gas emissions. Shared use of vehicles will reduce the number of cars on the road and thus congestion and emissions. Automation, in most cases, will reduce crashes. These benefits will be fully realized—indeed, enhanced—only if the three revolutions are integrated. Integration means EVs carrying multiple occupants under automated control. The result will be low-cost, low-carbon, equitable transportation.

Will this integration—the dream scenario—come about? As *Automotive News*, the principal trade magazine of the auto industry, put it, "There are millions of ways that flawed, messy, sometimes inconsiderate humans can mess up a perfectly good utopian scenario."[9] Many factors will shape the future, including how willing consumers are to accept new services and to share rides, how open transit operators are to embracing new mobility services as complements to their bus and rail services, and how inclined automakers are to become mobility companies instead of just car companies. Equally important will be the myriad policy, regulatory, and tax decisions made by local, state, and national governments—and how willing local politicians are to promote pooled services.

In the chapters that follow, we take a deep dive into the three revolutions and the potential synergies among them. We aim to inform and elevate the discussion about what needs to be done to unlock the enormous upside potential of vehicle electrification, automation, and ridesharing—and to avoid the downsides. Choices will be made by consumers, taxpayers, governments, transit operators, start-up mobility companies, and automotive and energy companies. With this book, we provide insight and knowledge that we hope will lead to wiser choices by all. We remain hopeful that more informed decisions will steer these revolutions toward the public interest and a better quality of life for everyone.

Overview of the Three Revolutions

Before looking at what might lie ahead, let's briefly take stock of where we are now with the three revolutions. We elaborate in the following chapters.

Vehicle Electrification

It's been a slow and circuitous journey. In 1900, about a quarter of all the cars in the United States were electric. They were quickly overtaken, though, by internal combustion vehicles and didn't mount a comeback until a hundred years later. In 1990, California took an important first step: it adopted a zero-emissions vehicle (ZEV) mandate intended to curb the air pollution blanketing Los Angeles. The mandate suffered a tortured life for two decades, as industry launched lawsuits and technology evolved more slowly than anticipated. General Motors (GM) leased a thousand sporty electric cars in the 1990s, to some acclaim, but crushed them a few years later.

The breakthrough came in 2008, building on decades of research and development, largely government funded, on batteries and power electronics. That year, Tesla jolted the automotive world with a racy, high-performance electric sports car and followed in 2012 with the sleek, elegant, powerful Model S sedan. In 2010, Nissan introduced its Leaf, the first mass-market EV in almost one hundred years, followed quickly by the plug-in hybrid Volt from GM. Observing these impressive technological advances, California rejuvenated its ZEV mandate in 2012, requiring automakers to ramp up EV sales to 15 percent market penetration by 2025. Nine other states followed suit—and as this book goes to print, China and the European Union are poised to follow with similar mandates.

China soon surpassed California in EV sales, with a half million electric cars, trucks, and buses sold in 2016.[10] China's motivation was to eradicate local air pollution and create a domestic motor vehicle industry that could leapfrog to global dominance. Cumulative global sales of all EVs—a broad category that includes those powered by batteries and by hydrogen fuel cells—reached two million in 2016.[11] Overall market penetration was still just 1 percent globally, but there were big success stories in smaller markets, such as Norway, which has only 5.2 million people but saw market penetration of EVs approach 35 percent in 2017.[12]

By 2017, every major automaker was making massive investments in EVs. More than thirty-five different models were for sale in the United States, and many more elsewhere, especially in China, with every automaker planning to expand its offerings. Battery costs were dropping faster than anyone had anticipated, and nations around the world were implementing aggressive policies in support of EVs. Where subsidies were massive, as in Norway and parts of China, sales were soaring, but where subsidies were more modest, progress was slower. In California, EVs captured 5 percent of the new-car market in 2016. But why weren't sales higher? In many regions of the state, consumers could find high-quality EVs for prices lower than comparable gasoline cars, thanks to generous incentives. The promise was great, but progress was measured.

Pooling and Sharing

The sharing economy and information technology have finally come to transportation, enabled by the release of the iPhone in 2007. Early champions Lyft and Uber first used smartphone apps to enlist private car owners in providing rides on demand. These initial offerings were glorified taxi services—glorified in the sense that they were superior to traditional taxi services in cost and convenience but still functioned like taxis. They were a first step toward transformational change.

Lyft's introduction of Lyft Line in 2014 was a truly game-changing event for sustainable transportation, making it possible for two strangers going in the same direction to seamlessly share the trip. Uber quickly copied Lyft with its own version of pooling, UberPool. Riders pay about half the normal price in exchange for sharing the car with other riders and accepting a detour of a few minutes to pick up and drop off a second (or third) passenger. By 2016, 50 percent of Lyft and Uber riders in San Francisco—where Lyft Line and UberPool were first launched—were opting to share rides with strangers.

The next frontier for Lyft, Uber, and similar companies is incorporating AVs. In 2016, Lyft partnered with GM to begin developing an on-demand network of AVs across the United States, with Lyft cofounder John Zimmer predicting that by 2021, the majority of Lyft rides will be provided by AVs.[13] In the same year, Uber launched its first fleet of technician-assisted self-driving cars on the streets of Pittsburgh and announced a partnership with Volvo to use AVs to offer shared rides.

But will large numbers of people be willing to share rides and get into robot cars with strangers? Given the small and dwindling number of conventional carpoolers, we should not assume people will forgo car ownership and readily share rides just because apps and smartphones make it easier.

Vehicle Automation

Self-driving cars once seemed like they belonged in some distant sci-fi future. First featured in the GM exhibit at the 1939 New York World's Fair and then demonstrated in the real world by GM and Honda in 1997, they are now nearing commercialization. In 2010, Google announced it had a car that was safely self-driving around San Francisco—with no special roadside infrastructure or city retrofitting. This was just six short years after not a single automated car had been able to complete a course in the middle of the desert set up by the Defense Advanced Research Projects Agency (DARPA), the US military agency that financed the Internet.

The reality is that most new cars in Europe, the United States, Korea, and Japan are already partly automated. Many have adaptive cruise control, which speeds them up or slows them down based on the speed of the car in front of them. Many also have emergency braking, whereby the car takes over control and stops when it detects an imminent crash. And many also have lane-keeping and blind-spot assistance, whereby the car detects when it is crossing a lane (without a turn signal on) and when another car is in the driver's blind spot. Much of the hardware and software needed to automate cars was already in place and commercial by 2015.

Today, a handful of companies are leading the way in developing AVs, with Tesla and Google (now Waymo) joining major automakers. By 2016, Tesla had sold fifty thousand vehicles to the public with Autopilot technology built in. Those cars were essentially capable of full autonomy, at least on freeways, where there are few obstructions and no traffic signals. Experience with partial automation and many test fleets gives companies like Tesla, GM, Ford, Toyota, BMW, Mercedes, Volvo, and Nissan the confidence to promise commercial sales of self-driving cars (but not driverless, as we explain in chapter 3) by 2020. Companies such as Uber say they will switch to automated cars as soon as they can in order to shed one of their most significant costs: drivers. There will also be big savings for the trucking industry, so it's no surprise that start-ups like Otto (founded by ex-Googlers and purchased by Uber in 2016) are testing long-haul trucks that drive themselves.

These hopeful visions are premised mostly on technical progress. But in major transitions like this, success often requires more than technological superiority. There will be regulatory battles over vehicle licensing, corporate battles as car companies transition from manufacturing to mobility services, and public debates over fascinating but remote dilemmas like self-driving cars being forced to choose between either holding their course and hitting Grandma or swerving into a troop of Boy Scouts.

The Road Ahead

There is much uncertainty about how these three revolutions will play out. We know for sure that cars are here for the foreseeable future and that most cars will eventually become electrified and automated. Once cars are automated and the driver has been removed from the car, an array of new uses will become economical. Some changes will happen more slowly than others. Some changes will be undeniably positive for consumers, the economy, and society, while others might be disastrous—in terms of overall cost, environmental impact, and social equity.

The Elusive Promise of Safer Vehicles and More Equitable Access

Robot cars will undoubtedly be far safer than ones piloted by humans— eventually. Robots won't drink and drive, won't get tired, won't be distracted by texting and children, and will have lightning-fast reflexes. In a world where all vehicles are self-driving, perhaps thirty thousand lives will be saved per year in the United States and millions of injuries avoided. But if humans are still driving or want to drive themselves, and if the transition from partially automated to fully driverless cars is delayed by safety regulators and local governments, as well as human reticence to cede control, we might not see changes in safety outcomes for many, many decades. In fact, deaths might increase if the technology is frozen one step short—with cars automated but requiring (unreliable) human intervention.

Likewise, the shared-use revolution, with the promise of low-cost travel, greater access for mobility-disadvantaged travelers, and less vehicle use, is also elusive. Personal cars are very expensive to own and operate, but most people have no alternative. Taking into account depreciation, insurance, fuel, registration, and maintenance, the full cost of owning and operating a new internal combustion engine car in the United States is about $0.57 per mile when the car is driven fifteen thousand miles per year. Used cars are less expensive, but not much, since older cars require more maintenance and more repairs. Where offered, bus and rail transit is much cheaper for travelers, but only because it is heavily subsidized. In the United States, riders pay only $0.25 per mile on average, but the true cost is five times that.

Travel by automated cars would be far cheaper if provided commercially, meaning the cars would be used more intensively. With the large up-front depreciation costs spread over many more miles, the cost could be reduced to about $0.20 per mile for an automated car logging one hundred thousand miles annually (with the vehicle assumed to cost $10,000 extra).[14] And if the rides were pooled, as with UberPool and Lyft Line, with no driver, the costs would be pushed down to $0.10 or less per passenger mile. These costs are summarized in figure 1.2.[15] This very low cost would assure more equitable access and also be highly disruptive in ways we are only beginning to imagine. But there's no guarantee that the pooling services being offered by Lyft, Uber, and others will flourish, especially outside dense urban areas. Indeed, the dominant shared vehicles in 2017 were still Uber and Lyft cars carrying one passenger at a time.

Congestion and Environmental Impacts: Down or Up?
The effects on traffic congestion could go either way. In the most positive scenario, if all AVs are shared rather than privately owned, the congestion problem evaporates. Vehicle use would drop significantly (thanks to

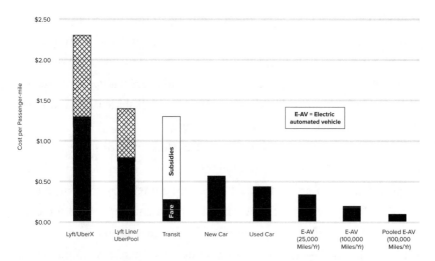

Figure 1.2. Comparative costs of travel by different means in the United States. Cross-hatched bars indicate the range of costs due to length of trips (longer is cheaper per mile) and variation across cities.

poolings), and road space utilization would improve dramatically. On-street and much off-street parking, including parking lots and garages, could be repurposed as public space—including wider sidewalks, more trees, bike lanes, and street furniture—and used for affordable housing and parks. On the other hand, if self-driving cars are privately owned by individuals, many of those expensive cars will spend considerable time circling the block endlessly and returning to remote parking lots instead of paying for parking. Think of what *this* will do for congestion. Similarly, some have suggested that private ownership of AVs will cause cities to sprawl into a new ring of "exurbs" as drivers forgo their distaste for car travel, in some cases abetted by being able to travel at higher speeds in AV-only lanes, though this sprawl could be restrained by land use rules, water and utility policies already in place, and people preferring urban lifestyles and walkable neighborhoods.

In terms of energy use and emissions, the potential synergies from combining the three revolutions are huge. Studies from independent groups such as the US Department of Energy,[16] the International Transport Forum,[17] the University of Washington,[18] and the University of Texas[19] suggest that shared, electric, automated vehicles could dramatically reduce greenhouse gas emissions. But all those studies are based on a few key assumptions. They assume travelers will share rides and make choices based solely on time and cost savings.

Perhaps. It's well documented that travelers and consumers do not act as purely rational economic beings. As for the vehicles, if AVs run on petroleum and are individually owned, sometimes cruising empty between drop-offs and pickups, emissions and energy use will increase. Figure 1.3 highlights the many ways in which automated cars could affect energy use and greenhouse gas emissions. Increases would be due to higher speeds, which greatly increase energy use, and increased travel, since people would not see time in a vehicle as wasted and because the young, the physically disabled, and others would be able to access vehicles more easily. Of course, if the vehicle were carrying multiple riders, the energy and greenhouse gas impact per trip (or per person mile) would be sharply reduced.

In a study of urban passenger travel worldwide, researchers at the University of California, Davis, estimated that with driverless cars but little pooling and electrification, greenhouse gas emissions would increase 50 percent and vehicle use 15 to 20 percent between now and 2050. In contrast, in a dream scenario where driverless cars are pooled and electrified, vehicle use would *drop* by 60 percent compared to business as usual, greenhouse gas emissions would drop by 80 percent, and overall costs of vehicles, fuel use, and infrastructure would drop by more than 40 percent—representing a savings of $5 *trillion* per year. Though these scenarios would play out in very different ways from one country to another, the researchers estimated that the scale of these reductions would be similar across most of the world.[20]

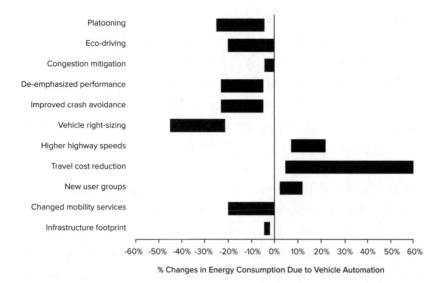

Figure 1.3. Projected changes in energy consumption due to vehicle automation. For each contributing factor, a range of values is estimated. Source: Zia Wadud, Don Mackenzie, and Paul N. Leiby, "Help or Hindrance? The Travel, Energy, and Carbon Impacts of Highly Automated Vehicles," *Transportation Research Part A* 86 (April 2016): 1–18.

Impacts on Road Financing and Jobs

Another uncertain and potentially problematic outcome is the issue of who pays for roads. The motor fuels tax, imposed by the federal government and states, was originally intended as a user fee for road construction and upkeep that people pay in rough proportion to the distance they drive. This simple and sound concept began to be undermined as vehicles became more efficient. Users of energy-efficient vehicles use much less fuel and thus pay much less per mile, while owners of fully electric vehicles do not pay any tax at all. This undermines federal and state funding for roads and transit; it also means that as more people purchase EVs, even less gas tax is paid. Because EVs today are disproportionately owned by higher-income households, and older cars, often gas guzzlers, are owned by less affluent people, the burden of paying for roads via the fuel tax is gradually being shifted onto lower-income drivers.

In our dream scenario of shared electric AVs, today's deteriorating user-based financing system completely collapses as vehicle use shrinks and most vehicles don't consume gasoline. This scenario is also a disaster for local governments, which depend on a mix of parking fees and fines, speeding tickets, and other car-related fees and fines to finance local roads, parking garages, and transit.

A different and better taxation system will need to be created—preferably one that imposes the full cost of car travel on drivers (and passengers), including not only the cost of building and maintaining roads but also pollution, congestion, sprawl, and climate change. Many observers envision a gradual, longer-term transition from fuel taxes to some form of mileage-based user fee that charges vehicles directly for their use of roads. Such fee systems raise their own set of equity considerations,[21] but they do lend themselves to refinements that take into account whether vehicles are used for shared multipassenger service and whether AVs are owned by individuals or commercial mobility providers.

One other pivotal concern is job loss. What of those 3.5 million freight and delivery truck drivers and 665,000 bus drivers in the United States? And the taxi and livery drivers—90,000 of them registered in New York City alone, not counting Uber and Lyft drivers? All these jobs are presumably at risk. But the story is more complicated. A massive shift to shared mobility, say 10 to 40 percent or more of travel, would convert a private, personal activity into a commercial activity. Many new jobs would be created for those managing the new businesses, cleaning and maintaining the vehicles, designing and updating the computer systems, dealing with disgruntled customers, and so on. Similarly, truck drivers might find new work managing aspects of automated freight hauling that haven't even been imagined yet. The job impact of pooling and automation is highly uncertain. With an integration of the three revolutions, the net effect may well be positive.

Steering toward the Public Interest

The three revolutions are part of larger societal and economic transformations. Globalization, industry automation, and part-time service jobs threaten social cohesion and have contributed to income inequality. Will the mobility sector be part of the problem or part of the solution?

Left to the market and individual choice, the likely outcome is more vehicles, more driving, and a slow transition to electric cars—reflecting many elements of our nightmare scenario. It is the tragedy of the commons, where no one is championing the public interest. Yes, it is possible to imagine a sustainable AV future without pooling premised on individual ownership—at least in rich countries such as the United States—but only if the vehicles are electrified, road pricing is enacted judiciously, conventional transit continues to be funded, and transport finance is transformed so that those with low incomes and physical disabilities still have access to work, school, health care, and other activities. But that is a lot of "ifs." If all these contingencies don't materialize, the alternative to pooling is the nightmare scenario.

The future is likely to be far more positive if government intervenes and guides the invisible hand of the marketplace to avoid the excesses of self-interested behaviors. Antitrust laws, along with regulations focused on pollution prevention and worker protections, are all aimed at minimizing the direst excesses of capitalism. In a dream scenario, they guide powerful stakeholders toward the public interest.

That is the challenge here. For the first time in many decades, mobility is on the cusp of not just one transformation but several. The dream scenario depends critically on the merging of electrification, pooling, and automation.

In practice, how does this dream scenario unfold? The key first step seems to be the creation of travel choices that increase convenience and comfort and reduce personal travel costs. Not until people have choices will they be willing to give up their personal cars—the linchpin for a cascade of changes. And not until they have choices will they accept pricing policies—policies that encourage pooling and discourage ownership of personal vehicles. Pooling and pricing are key to setting us on a path to less costly, less resource-intensive, more widely accessible, and more sustainable transportation.

Change is going to happen. And it will be transformational. Many will try to slow it, while others will try to accelerate it. We can be ostriches with buried heads and hope that disinterest, disengagement, and the normal workings of the marketplace will somehow turn out well. Or we can apply our best thinking to harness vehicle electrification, mobility sharing, and automation to create better cities, a livable planet, and a future that serves us all. Read on.

Key Policy Goals

These are the overarching goals in transitioning to a safer and more economical, environmentally benign, and equitable transportation future:

- **User incentives:** Incentivize travelers to choose pooled mobility services over individual ownership of vehicles now, and more so when driverless vehicles become available.
- **Pooling and EVs:** Encourage mobility service companies to embrace pooling and EVs (including hydrogen fuel cell vehicles). Motivate automakers to design AVs for pooling that are powered by electricity or hydrogen.
- **Equity and transit:** Encourage transit operators and mobility service companies to collaborate in providing more access and service at lower cost.
- **Land use:** Begin redesigning cities for a transportation system that uses less parking and fewer roads and is more conducive to pooling, walking, biking, and affordable living.

In the following chapters, we elaborate on specific policies that could be adopted to achieve these broad goals.

CHAPTER 2

Electric Vehicles: Approaching the Tipping Point

Daniel Sperling

Electric vehicles have come far and will eventually dominate—even in the realm of automated and shared vehicles—but progress will be slow without strong policies.

IN THE LAST THREE DECADES, ELECTRIC VEHICLES (EVs) have vastly improved in every way—in cost, performance, efficiency, and availability to consumers. My own history with EVs shows how much.

My first drive in an electric car was in 1992 in a Geo Metro, retrofitted by James Worden of Solectria. It topped out at sixty miles per hour, whined loudly, and quit after about forty miles. Then I bought a 2011 Nissan Leaf. It wasn't the prettiest or fastest car on the road, but it was soothingly quiet, accelerated smoothly, and heated and cooled me comfortably. Its one shortcoming was driving range, advertised at one hundred miles. Alas, as my wife and I virtuously drove our new Leaf away from the dealership, we watched in despair as the one-hundred-mile range displayed on our instrument panel slipped to eighty miles as we accelerated onto the freeway. After dropping me off at work, Sandy realized she couldn't make it home that first day. She found her way back to the dealer for a two-hour charge to get home. Too many others were painfully initiated in the same way into the reality of range anxiety and slow charging.

Our next EV, a Tesla Model S, swept away the shortcomings of the Leaf with its range of 270 real-world miles. Its huge smart screen would tell me exactly where the next supercharger was, how to get there, how long it would take, and how much range would be left when I arrived. The charge was free, and it took thirty to forty-five minutes to recover 80 percent of my range. There was only one downside: the car cost more than $70,000.

My current EV is a Mirai, a hydrogen fuel cell car made by Toyota. It also is zero emitting, with the additional virtues of being priced 50 percent lower than the Tesla, having a somewhat longer range, and refueling in four minutes at a hydrogen station. Unfortunately, there is only one hydrogen station in the entire Davis-Sacramento area where I live; fortunately, the station is on the way to my most common destinations, downtown Sacramento and the Sacramento airport. Already there are enough strategically located stations in California for me to go skiing at my favorite resort, visit my wife's family three hundred miles away, and drive to Los Angeles, where numerous stations await me. The GPS in my car and the apps tell me exactly where the stations are and how to get there.

For the first time in their history, EVs are finally the equal of gasoline cars in performance, styling, and price (though not yet cost, as explained later). And they are superior in quietness and driving experience for most drivers, according to surveys. Every major automaker offers a variety of EVs for sale, often at attractive prices, and more and more are selling hydrogen fuel cell EVs. A growing number of analysts foresee demand for EVs accelerating sharply in the coming years as costs continue to fall, driving ranges increase, and governments become more insistent.

The Current Reality

For now, EVs are considerably more expensive to produce than gasoline cars, even with dropping battery costs. Automakers price them lower than their cost in order to sell them, to satisfy government mandates in some areas, and to help meet aggressive fuel efficiency and greenhouse gas standards. The price tipping point for small EVs will likely be sometime in the 2020s as battery costs continue to drop and vehicle efficiency and greenhouse gas standards in China, Europe, and the United States start to push automakers beyond gasoline and diesel.

The transition is just beginning. Market share is higher where market support is especially strong—where there are lots of incentives and electric charging stations and where government leaders tout the future dominance of EVs. The largest single market is China, with half of the world's 1 million EV sales in 2016—including more than 320,000 cars, 120,000 buses, and 60,000 medium- and heavy-duty trucks.[1] About one-fifth of total sales were in the United States. The country with the highest market share was Norway, where EVs accounted for nearly 35 percent of vehicle sales by July 2017—but Norway has only 5.2 million people. Overall, EVs accounted for only 1 percent of total vehicle sales in the world in 2016.

While some of us are highly knowledgeable about EV offerings and the various incentives from governments, most are not. In a series of consumer EV studies, Ken Kurani of the University of California, Davis, found that car owners were remarkably ignorant of EVs, even in California.[2] For instance, less than 2.5 percent of his sample of new car buyers in California through December 2014 reported extensive driving experience with EVs, and only 10 percent reported that they had anything more than cursory experience—this in a state that had been aggressively promoting EVs for twenty-five years, had a wide range of models available at dealerships, and offered significant incentives through which local electric utilities had been promoting subsidized home charging.

The same survey found that only 7 to 8 percent of the households that shopped for or bought new vehicles in 2015 in California, Oregon, and the northeastern states actively shopped for or bought an EV. And in Germany, where the government enacted an EV subsidy in July 2016, the subsidies were sitting largely unused by the end of the year, with EVs capturing only half of 1 percent of sales.[3]

One other number shows how far EVs have to go: Tesla delivered 76,230 EVs in 2016, up 51 percent from a year earlier, but in comparison, Ford was selling the same number of F-series pickups every month.[4]

Although technology challenges remain, the bigger challenge is now consumers—how to motivate them to embrace battery EVs, as well as plug-in hybrid and fuel cell EVs. But before exploring how to influence consumer behavior, let's explore the larger context of EVs—why nearly everyone in the industry considers EVs inevitable.

Why Does the World Need Electric Vehicles?
The modern embrace of EVs is fairly recent. In my 1988 book, *New Transportation Fuels*, I barely mentioned EVs. At that time, climate change was just one of several concerns, along with local air pollution, economic impacts, and energy security. Many compelling substitutes for petroleum were under consideration. However, by the time of my 1995 book, *Future Drive*, I was arguing that electric propulsion was the key to sustainable transportation and energy, and by 2017, the entire automotive industry had come to embrace EVs as inevitable and desirable.

Four reasons account for this growing embrace of EVs. The first is environmental impact. In many polluted cities—in the rich industrialized world as well as across Asia and increasingly in Latin America and Africa—the most compelling argument for EVs is unhealthy (and unsightly) air. Indeed, worsening urban air pollution is central to China's commitment to EVs.[5] And besides causing local air pollution, motor vehicles emit about 20 percent of all greenhouse gas worldwide.[6] EVs represent the best hope for dramatic reductions in transportation emissions. Pure battery EVs and hydrogen fuel cell EVs emit zero greenhouse gases and zero pollutants during driving, while plug-in hybrid electric vehicles (PHEVs) have zero emissions while operating on electricity.

The Committee on Climate Change, an official advisor to the UK government, says that to cut carbon emissions at least cost, 60 percent of new car sales in the United Kingdom should be electric by 2030.[7] Others go further. In May 2016, the head of India's Power Ministry announced a goal that all vehicles on India's roads would be electric by 2030.[8] And in October 2016, Germany's national legislature passed a resolution to ban the internal combustion engine starting in 2030 and allow only "zero-emission passenger vehicles" to be approved.[9] While these announcements were more aspiration than reality, they signify an embrace of EVs as one of the best ways to cut pollution and reduce global warming. As we will see, these aspirations are increasingly being converted into aggressive government regulations and ambitious incentives.

It should be acknowledged that although the energy and pollution advantages of EVs can be large, they depend on the source of the electricity used to charge the vehicles. EVs provide the greatest environmental benefits when electricity is generated from renewable energy, including wind and solar, as well as nuclear and hydropower. When electricity generation is switched from fossil to renewable and nuclear energy, the climate benefits of EVs are huge—almost a 100 percent reduction in greenhouse gas emissions. In California, more than half of the electricity already comes from zero-emitting wind, solar, nuclear, and hydropower sources. In this case, battery EVs provide huge improvements over gasoline and diesel vehicles, measured on a life-cycle basis. Likewise, in France, where most electricity comes from nuclear power, the environmental benefits are enormous.

EVs are less attractive where most of the electricity comes from coal, such as in China and India, and to a lesser extent in parts of Germany and the United States. In these cases, there are still local pollution benefits because actual exposure to the pollution is limited (with power plant emissions often in less populated, remote areas). Perhaps more important, the electricity grid is being decarbonized virtually everywhere. Around 2010, Germany, the United States, China, and many other countries began aggressive initiatives to increase the use of renewable energy to generate electricity. As the world decarbonizes electricity generation and eventually hydrogen production, EVs become even more attractive environmentally.

A second large attraction of EVs is technological. Although incumbent automakers are anchored by the legacy of one hundred years of internal combustion engines and mechanical engineering designs, they are well along in converting the infrastructure of the car to electronic controls. Since the mid-1970s, they have slowly embraced the use of electricity and electric motors to manage everything from opening and closing windows to steering, braking, and accelerating. The advantages of an EV architecture are reduced cost and weight, greater reliability, easier maintenance as a result of fewer moving parts, and more precision in braking, managing combustion, and shifting gears. Hydraulic braking with fluids and hoses is disappearing, as are steering wheels connected to long, heavy steel rods. Electric motors are proliferating to control sunroofs, windows, seats, and much more.

Many of these electronic and digital attractions carry over to electric propulsion. Not only do vehicles benefit from the replacement of transmissions, mechanical drivelines, and heavy combustion engines by electric motors and components that are physically lighter and easier to manage, but they also benefit from the ability to capture energy from braking. Vehicle electrification represents a huge advance in energy efficiency.

Third, electrification opens up design opportunities. With modular batteries that can be placed in and under the frame, the car is more stable, allowing more opportunities to enhance performance. And when the car has neither a metal transmission and driveline through the middle of it nor a radiator and engine block in the front, it can be completely redesigned to better serve passengers, with advantages including easier ingress and egress.

This design flexibility also allows automakers to use just a few vehicle platforms—perhaps just three for small, medium, and large vehicles—as opposed to the ten or twenty platforms needed by automakers for their mix of more complex combustion engine vehicles. Cost savings in scale economies and simplified manufacturing from reducing the number of platforms could be massive.

Fourth, vehicle electrification is good for consumers. EVs reduce dependence on foreign oil and consumer vulnerability to volatile fuel prices. EVs are much more energy efficient than conventional vehicles, so they generally cost less to operate. This efficiency will steadily improve as components are improved, parasitic losses are reduced, and overall energy management is optimized. EVs are also quieter and drive more smoothly than combustion engine vehicles. They are less expensive and easier to maintain because they have fewer moving parts and no need for oil changes. The possibility of home recharging is also attractive to most consumers. Plugging in will be difficult for apartment dwellers and some homeowners, but it's a comfortable experience for most people and a preferred option for many.[10]

While it's clear that EVs are in the public interest and inevitable, what is less clear is how fast the transition will occur and how it might be accelerated to capture EVs' many benefits as quickly as possible. The train has left the station on electrification, and automakers and governments are aligned in their commitment to EVs, but electrification is still in its infancy and faces some stiff resistance. A review of the history of EVs highlights the barriers and the instrumental role of government regulation in getting them to where they are today and where they will be in the future.

The Long Slog to Market Ready

Although my personal history with EVs goes back to 1992, EV technology goes back more than a hundred years. True to what Steven Poole asserts in his popular 2016 book, *Rethink: The Surprising History of New Ideas*, EVs emerged not in a flash of inspiration but rather in a long slog of incremental innovations, persistence, and increasing investments and focus by researchers, government, entrepreneurs, and big businesses. Battery technology held back progress for more than a century. But after 2010, government fuel economy and emissions standards began to force the auto industry toward electrification, resulting in major new investments and rapid innovation. By 2017, major cost and range breakthroughs meant that EVs were poised to capture an expanding market.

Early Electric Vehicles: Overpowered by Combustion Engines

EVs originated in Europe. Robert Anderson of Scotland is credited with inventing the first crude electric carriage, powered by nonrechargeable primary cells, sometime between 1832 and 1839, and another Scotsman, Robert Davidson, made a model electric locomotive in 1837 powered by zinc-acid batteries.[11] The rechargeable lead-acid battery was invented in 1859 by French physicist Gaston Planté. The first electric bus appeared on the streets of London in 1886, just a year after the Germans Gottlieb Daimler and Karl Benz built the first gasoline-powered vehicle.

EVs outnumbered gasoline-powered vehicles at the turn of the century.[12] They were attractive because they offered superior ease of starting and driving. "The electric runabout . . . appears to be the most popular form of automobile for women, at any rate in the National Capital," noted a society columnist in 1904 (see figure 2.1). W. C. Durant, the founder of General Motors (GM), pointed out that gasoline-powered cars were "noisy and smelly, and frighten the horses."

But women did not buy many vehicles, and battery technology did not improve nearly as fast as gasoline vehicle technology. By 1910, the

Figure 2.1. A 1906 Columbia Mark 68 Victoria electric car and a 1912 hand-cranked battery charger, both made by Pope Manufacturing Company. Source: Museum of Innovation and Science, Schenectady, New York.

heyday of the electric runabout was over. The Ford Model T was now selling for less than half the price of any advertised electric car. By 1915, less than 2 percent of the 2.5 million motor vehicles in operation in the United States were powered by electricity. The EV industry dwindled away, with the last electric car factories in the United States closing in 1935. The only EVs that continued to be produced in the United States were forklifts and other equipment that operated indoors and did not require much range.

The early electric cars ultimately failed because of poor performance and high battery costs—and rapid improvements in internal combustion engine vehicles. Charles Kettering's electric starter overcame the need to hand crank, engines were made more efficient, the use of rubber engine mounts reduced vibration, and advances in carburetion and ignition made gasoline cars easier to drive. Meanwhile, oil discoveries swept away energy supply concerns, and a spreading network of roads rewarded the much greater range of gasoline cars.

Here matters stood for decades. The invention of semiconductors in the 1950s and continuing improvements in motors and controllers spurred some interest in EVs in the 1960s. GM and Ford initiated modest research programs on EVs, with GM displaying a converted Corvair running on silver-zinc batteries at the 1966 auto shows. In Europe, small numbers of EVs were used as delivery vehicles. British manufacturers never stopped producing EVs, turning out several thousand milk delivery trucks a year, with an estimated twenty-five thousand still in use in 1993.

In Japan, all the major automakers embarked on determined EV development programs in the 1960s, motivated by Japan's dependence on imported oil, with at least four major companies building advanced prototypes that were reportedly ready for commercial production in the 1970s. But they held production plans in abeyance and apparently cut back their R&D efforts during the 1980s, satisfying themselves with building only a few very small EVs for their local market. They were reluctant to market EVs more broadly, sharing the concerns of the three US automakers that EVs were too costly and risky.

The First Electric Vehicle Comeback and
California's Zero-Emission Vehicle Mandate
The first EV comeback was in 1990, when GM and Aerovironment unveiled a sleek, fast, electric sports car called the Impact. It was a prototype, not ready for commercialization, but it came at just the right time for California regulators. California was suffering from severe air pollution. Much progress had been made—engine emissions were much lower thanks to the widespread use of catalytic converters—but air pollution in Los Angeles and vast parts of the interior of the state was still very unhealthy. With expanding car use, something dramatic was needed.

The California Air Resources Board (CARB) was in the process of adopting more stringent, "ultralow" emissions standards for cars. Inspired by GM's electric prototype, the board quickly added a rule, essentially a footnote to the broader emissions standards, requiring automakers to start phasing in zero-emission vehicles (ZEVs)—amounting to 2 percent of their sales in the state by 1998 and 10 percent by 2003.[13] This ZEV mandate would eventually play a central role in the transition to EVs, but only after a far more protracted and torturous journey than regulators anticipated in 1990.

CARB is a relatively independent board responsible for administering California's air pollution and greenhouse gas laws. Because the state suffered unusually severe air quality problems, the US Congress in 1967 granted California authority over vehicle emissions, as long as its rules are at least as strong as the federal ones. Other states were subsequently given the option of following the more stringent California standards; nine states had chosen to adopt California's ZEV mandate as of 2017 (and several more had also adopted California's vehicle emissions standards). Thus California has positioned itself in a leadership role ahead of the federal government, and CARB has become the world's preeminent air quality agency, pioneering emissions requirements for vehicles and fuels, first premised on local air pollution and now also premised on greenhouse gas impacts.[14] The ZEV mandate was and still remains a central pillar in the state's campaign against air pollution and climate change.

When CARB issued its 10-percent-by-2003 ZEV requirement in 1990, it grabbed the attention of automakers—and the public. It triggered a wave of investment in EVs and batteries. It motivated the US government in 1991 to launch the Advanced Battery Consortium, a large applied R&D program intended to help US industry leap beyond lead-acid batteries to the next generation of more powerful and compact batteries—and governments in Japan and Europe to launch similar, but smaller, battery programs.

But the ZEV mandate of 1990 turned out to be far too aggressive. Batteries did not improve as fast as regulators hoped, and consumers were reluctant to embrace these new types of vehicles. The mandate was whipsawed by industry lawsuits on the one hand and EV advocates on the other. The auto industry, almost with a single voice, complained that California was forcing technology that was not yet ready into the marketplace. On the other side of the fence were those who argued that the ZEV program was necessary to accelerate the development and commercialization of advanced vehicle technologies.

I saw firsthand what was at stake when I testified before CARB at a public hearing in May 1994. Buses of demonstrators who were ostensibly members of a grassroots lobby, Californians Against Hidden Taxes (CAHT), rolled in to denigrate EVs. It turned out that the group was heavily, perhaps totally, subsidized by the oil industry. Soon referred to as Astroturf groups because the local roots were artificial, they gave the appearance of massive popular opposition.

GM's electric prototype went into production as the EV1 in 1996 (see figure 2.2). It encapsulated both the challenge and the opportunity facing EVs. On the one hand, the car was incredibly fast for its time, quietly racing from a standstill to sixty miles per hour in eight seconds; but on the other hand, the small car carried an astounding 1,175 pounds of lead-acid batteries, nearly 40 percent of its total weight. Those batteries lasted only two years, provided only 16.5 kWh of storage capacity, and propelled the vehicle just sixty miles on a charge.[15] GM produced only 1,117 of the cars, eventually shutting down production six years later in 2002.[16] It recalled all EV1s (which had been leased) and crushed all but a few, ostensibly to absolve itself of any liability and responsibility. EV advocates were outraged.

All told, automakers supplied only about two thousand electric cars in the United States in the late 1990s and early 2000s, most of them in California.[17] Faced with strong opposition by the auto and oil industries, CARB weakened the ZEV mandate in 2003, giving automakers

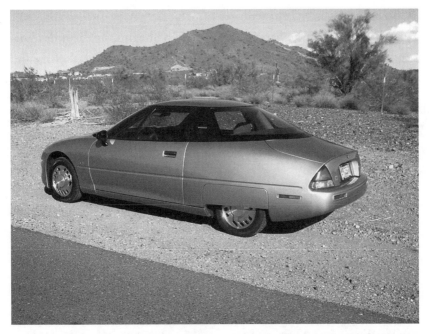

Figure 2.2. The EV1, stylish in its day. Photo by RightBrainPhotography (Rick Rowen), CC BY-SA 2.0 (http://creativecommons.org/licenses/by-sa/2.0), via Wikimedia Commons.

the option of selling a few hundred fuel cell vehicles through 2007 as a substitute for meeting battery EV requirements.

The 2006 documentary *Who Killed the Electric Car?* sought a culprit for the demise of the EV1. The film fingered GM, of course, but also the oil industry for opposing EVs, as well as the George W. Bush administration for opposing the ZEV mandate, consumers for not buying the cars, and CARB for weakening the ZEV mandate and embracing hydrogen fuel cell cars. It's true that the car companies never made much effort to market the vehicles, having convinced themselves that the cost was too high and the market too small. It's also true that the oil companies waged a fierce battle to defeat the electric-powered auto. But missing from the lineup was the one real culprit: the battery.

Battery Blues—and Breakthroughs

Batteries had been the bane of EVs since the beginning. Thomas Edison launched an intensive effort in the 1890s and devoted years and millions of dollars to finding a better battery. He aimed to improve on the large weight and short life of lead-acid batteries. His team conducted thousands of tests on all sorts of metals and other materials, finally settling in 1903 on an alkaline battery with iron and nickel electrodes. With great fanfare, it was introduced commercially. But when put to use, the batteries failed quickly and were not significantly better than lead-acid batteries. Edison shuttered his factory for three years to redesign the battery, but by 1910, the Model T was on its way to vanquishing EVs. The market for EV batteries was disappearing. Edison had failed, and with him—for the time being—the EV.

Better batteries were needed. After decades of slow progress, battery chemistry innovation began to accelerate in the 1980s, motivated by the proliferating market first for battery-powered consumer goods and later for cars. Companies around the world experimented with and eventually commercialized a variety of battery chemistries. In the early 1990s, Ford invested in and used high-temperature sodium-sulfur batteries in its electric pickup trucks. Others invested in nickel-cadmium and sodium-nickel-chloride batteries. All failed to measure up due to either toxicity or short battery life.

The first big breakthrough came during the latter part of the 1990s with major advances in nickel-metal-hydride batteries. This new battery technology was invented in the 1960s but was not widely commercialized until Stan Ovshinsky, the celebrated self-taught scientist, developed a more advanced alloy structure that was used in the last four hundred EV1s and in the first gasoline-electric hybrids—the Honda Insight launched in 1999 and the Toyota Prius in 2000. Then a decade later, a still better set of batteries, based in lithium, swept into the automotive world (though Nissan had been using them in demonstration vehicles in the 1990s). The new advanced batteries had far higher energy density (energy per unit volume), used low-cost materials, and were durable. Battery costs began a steep decline.

Costs will continue to fall and the scale of battery production will ramp up considerably as a number of large new factories come online, including Tesla's Gigafactory in Nevada. Built to supply enough batteries to support the car manufacturer's projected vehicle demand and burgeoning electricity storage business, Tesla's factory by itself will produce more lithium-ion batteries annually than were produced worldwide in 2013. In cooperation with Panasonic and other strategic partners, Tesla hopes to drive down the cost of its battery pack through economies of scale, innovative manufacturing, reduction of waste, and concentration of manufacturing processes under one roof.

At the same time, research on battery chemistry continues. Steve LeVine, author of *The Powerhouse: Inside the Invention of a Battery to Save the World*, calls the quest for a better battery "one of the single most important engineering and scientific pursuits currently going on. It's the Holy Grail."[18] Still, the continuing (significant) reductions in battery costs through 2030 and beyond are expected to come from incremental refinements in lithium-ion batteries, not new battery chemistries.

Tesla: Making the Electric Vehicle an "Object of Desire"
Tesla resuscitated the EV. After early management struggles, Elon Musk took control of the company and in 2008 unveiled a racy sports coupe— following in the footsteps of the aborted EV1. The Tesla Roadster was so powerful that it outraced almost all gasoline sports cars. It was powered by lithium-ion batteries, as were almost all EVs that followed (except small, low-cost EVs in China). The Roadster quickly captured the imagination of the media and other automakers.

Bob Lutz, vice chair of GM, famously asserted in an interview with *Newsweek* magazine that if a Silicon Valley start-up could build such a compelling car, so could GM.[19] The GM product turned out to be the Chevy Volt, released in December 2010. It was nothing like the Tesla Roadster, but it was an engineering marvel. It combined a sizable battery (16 kWh capacity) with a conventional gasoline engine. The car could drive thirty-eight miles on a charge and then switch to hybrid mode, with the battery and the engine working seamlessly together.

Nissan was another pioneer. It unveiled the Leaf, a pure battery EV, just before GM's plug-in hybrid Volt. The car was the first mass-marketed pure EV in almost a hundred years. It had a rated range of one hundred miles, but as my wife lamented, that was downhill with a tailwind. In normal driving, it achieved seventy to eighty miles. In cold weather or at freeway speeds, the range was substantially less. It worked for early adopters willing to suffer inconveniences—such as turning off the air conditioner on hot days—and those who made only short local trips. According to simplistic studies of drivers, it was good enough for more than 80 percent of all travel days. But drivers had become accustomed to the freedom of driving wherever and whenever. Range anxiety was a palpable concern for most drivers—including me.

In 2012, Tesla again became the hero, unveiling its Model S sedan. It revolutionized the industry and the market. *Consumer Reports* called it the best car it had ever tested, matched only by one Lexus model some years earlier. *Car and Driver* said, "It dispels conventional thinking about EVs—it's a glimpse of the future." *Alberta Oil* magazine had this to say: "The Tesla Model S is one of the most beautiful and interesting automobiles to ever get made. It might also be one of the most dangerous. That's because it's managed to do something that no other electric vehicle has ever achieved: become an object of desire."[20]

It was sleek and elegant (see figure 2.3), with a raft of innovations and a range of about 270 miles. It could carry up to seven people when an optional rear seat was installed, something Elon Musk had insisted on because he wanted a car that could carry his entire family. Yet this

Figure 2.3. The Tesla Model S, making EVs an "object of desire." Photo by Mariordo (Mario Roberto Durán Ortiz), CC BY-SA 4.0 (http://creativecommons.org/licenses/by-sa/4.0), via Wikimedia Commons.

large family sedan had the same extraordinary speed of the Roadster and was still able to outrace Porsche and Ferrari two-seat sports cars, leaping from zero to sixty miles per hour in less than four seconds. Innovations included a large flat screen that replaced all dials and knobs on the dashboard. Periodically, the map data and algorithms for the GPS and everything else were updated via the Internet—something never before done by any automaker and still not matched as this book is written. The only drawback was its $70,000 price tag.

Between 2010 and 2016, automakers introduced more than twenty-four EV models and sold more than four hundred thousand new EVs in the United States, with half of those sales in California.[21] Worldwide, by 2016, the largest EV supplier was BYD of China, followed by Tesla, BMW, Nissan, and two more Chinese companies ranked fifth and sixth. Of these EVs, about half were pure battery types and the other half were plug-in hybrids. Only about a thousand fuel cell EVs were sold, all in 2016. Even with the low market penetration rates of EVs (1 percent of global car sales), the batteries used in those vehicles were already exceeding the total number of batteries sold for all other applications (measured as kWh).

Then at the end of 2016, GM crossed a significant threshold—rolling out the Bolt, an EV with a driving range exceeding two hundred miles at a price less than $40,000, roughly equal to the average price of a gasoline car. Tesla followed a year later with its Model 3, with similar range and cost. It had garnered 325,000 preorders within a week of being announced,[22] sending a signal of pent-up demand.

The Zero-Emission Vehicle Mandate: Still Alive and Kicking

Analysts have observed that the EV market in the United States is pushed more by California's ZEV mandate than by federal regulations. Indeed, the ZEV mandate has continued to play a central role in the transition to EVs.

After weakening the mandate in 2003, CARB finally held the line in 2008. I had been appointed to CARB by Governor Arnold Schwarzenegger in 2007 and played a central role in the deliberations. The new rule required automakers to produce a total of 7,500 fuel cell vehicles, 12,500 battery EVs, or some combination thereof between 2012 and 2014, along with 58,000 plug-in hybrids.[23] This was considered far too aggressive by cautious automakers and far too weak by strident EV supporters.

Between 2008 and 2012, the debate flipped. With the bankruptcy and government bailout of GM and Chrysler, shrinking battery costs, growing concern for climate change, and a growing recognition of the positive attractions of EVs, the auto industry warmed to EVs. With my support, CARB proposed an aggressive strengthening of the ZEV mandate, requiring automakers to produce approximately 1.5 million EVs by 2025, representing about 15 percent of all sales in that year. Only Volkswagen opposed the new requirement. It was adopted by CARB in January 2012, with automakers quibbling over minor details but widely supporting the new rules. I recall two different senior executives telling me of their joy that for the first time, they were not party to their company's obstructionism but now part of the solution.

The path forward is becoming less controversial. One indication of automotive companies embracing vehicle electrification occurred at a small invitation-only workshop I hosted in September 2016 to address possible changes in California's ZEV mandate, attended by government relations directors from all the major car companies, as well as government regulatory leaders and NGOs. Not a single automotive executive suggested that the 2025 requirements for the ZEV mandate should be weakened or rolled back in any way (though at that time, Hillary Clinton was expected to win the US presidential election). The entire discussion centered on changes to strengthen and improve the mandate.

This unity of purpose began to fray after the election of Donald Trump as US president, with his appointment of senior government officials skeptical of climate change and government intervention. Automakers, through their trade associations, began to question California EV regulations and federal emissions standards. Soon after the election, Ford's then CEO Mark Fields said the automaker planned to lobby the new president to soften US and state vehicle standards, which he claimed hurt profits by forcing them to build more electric cars and hybrids than were warranted by consumer demand. At the same time, Fields said Ford had made no change in its plan to invest an additional $4.5 billion in hybrids and EVs by 2020 and to install these technologies in 40 percent of its nameplates.[24]

It is widely accepted that at some point, economies of scale and declining costs will kick in, and the market that was jump-started by California's regulations will take on a life of its own through normal market competition, consumer demand, and broader greenhouse gas performance standards for vehicles. For the United States and Europe, that is probably well after 2020. For China, it might be sooner.

A Slow Transition Ahead

Many questions remain about the future of EVs, including the likely mix of pure battery EVs, PHEVs (like the Chevrolet Volt), and fuel cell EVs powered by hydrogen. But what is almost certainly true is that EVs will eventually sweep gasoline and diesel cars from the market; it's a question of when. Such an assertion would have terrified the conservative auto industry of the 1990s. At that time, the industry and the technology were not ready. Having to meet a sales target of 10 percent EVs in 2003 would have been highly disruptive.

Now the industry is ready. The technology is sophisticated and efficient, automakers have the know-how, and the component supply chains are in place. The shift from mechanical engineering to electrical and computer engineering, already under way, will continue. Sales of the new electric propulsion technologies will accelerate. It won't be easy. Companies still need to bring down costs and improve energy efficiency, and they need to gradually shift their business models away from a financial and cultural embrace of internal combustion. But this transition no longer threatens the core business of automakers and is unlikely to be highly disruptive.

The transition to electrification of vehicles has been gradual and will probably continue to be so for three primary reasons: consumer caution when it comes to large purchases, the high initial cost of manufacturing EVs, and pushback from vested interests. There will be special circumstances, as in Norway, where rapid transitions are possible, but in places such as the United States—where the automotive market is large and diverse and cities, states, electric utilities, and others are intervening in differing ways—the overall pace will be far slower.

The International Energy Agency scenarios in figure 2.4 give one indication of what the range of future EV sales volumes might be. The high-volume scenario reflects a future in which the world is committed to keeping global temperature increases to two degrees Celsius. The Paris Accord scenario reflects what countries agreed to in the 2015 Paris Accord, and the reference scenario is what is likely without new government policies or significant shifts in EV market demand.

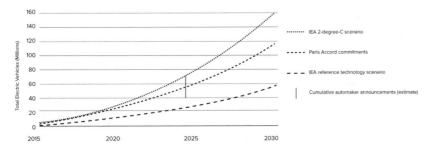

Figure 2.4. Deployment scenarios for the global stock of electric cars to 2030. Source: Adapted from International Energy Agency, *Global EV Outlook 2017* (Paris, France: OECD/IEA, 2017).

Driving Down Costs

Cost reductions will take time. Gasoline cars have benefited from a century of intensive development, while electric cars have been the focus of major manufacturers since only about 2010. Had EVs and their large battery packs also benefited from a hundred years of intensive development, the EV story might be quite different. As it is, pure battery electric cars cost about $10,000 more to manufacture than comparable gasoline cars.

Battery costs are still the critical variable. While shrinking rapidly, battery costs will remain substantial into the foreseeable future. Figure 2.5 shows battery costs falling from more than $1,200 per kWh in 2005 to as low as $200 per kWh for some suppliers by 2020. But even with this 85 percent drop, the cost premium is still large; for the Chevy Bolt, with more than two hundred miles of range, its 60 kWh of batteries would still cost $12,000 in 2020. When the cost falls to $150 per kWh, the Bolt batteries will cost $9,000. This additional cost is partly offset by the removal of combustion engines, transmissions, and pollution control equipment. With continuing cost reductions, it is plausible and even likely that by 2025, small EVs will become cost competitive on a total-cost-of-ownership basis.[25]

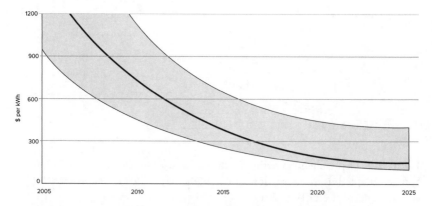

Figure 2.5. Battery costs are dropping sharply (the shaded area shows the possible range of costs), and with them, the costs of EVs. Source: Adapted from Björn Nykvist and Måns Nilsson, "Rapidly Falling Costs of Battery Packs for Electric Vehicles," *Nature Climate Change* 5 (2015): 329–32.

Energy costs also play a role and are illustrative of the challenge presented by the move to EVs. The widespread assumption is that energy costs are lower for EVs, but that is not true for many current EV owners. That is because many homeowners are subject to tiered electricity rates—the more you use, the higher the cost per kWh of electricity. The extra electricity for the car pushes many homeowners into a higher tier. They often end up paying more for electricity than they would have for gasoline.

Electric utilities and their regulators originally embraced tiered rates in many states as a way of motivating energy-efficient behavior. When people pay higher rates for using more electricity, they buy more efficient refrigerators and turn off lights. But adding at-home charging is penalized by that pricing logic. The solution is for electric utilities to adopt new rate structures to accommodate and support the introduction of EVs. It is in their interest to do so because EV charging provides them a mechanism for managing their electricity loads more easily and cheaply. But it represents a radical departure for many utilities and their regulators, and thus it will happen only gradually.

The new, more desirable approach will be to incentivize EV charging during times when demand is otherwise low, thereby allowing utilities to flatten the demand peak and spread the cost of their power plants and infrastructure over a larger base. If they can precisely manage when EVs are charged—by turning electricity to EVs on and off depending on when the wind blows and the sun shines and when customers ratchet their consumption up or down—they can treat the EVs as a massive storage pool. Some electric utilities already provide time-of-day rates that, in a crude way, anticipate shifts in demand and supply. They will increasingly use the smart meters being installed in businesses and homes to charge EVs when supply exceeds capacity and vice versa. Eventually, electric utilities will further integrate EVs into the grid by sucking electricity from EV batteries when more electricity is needed to avoid brownouts and blackouts and supplying electricity to the cars when there is excess in the grid. When all these changes happen, the cost of electricity to EV owners will drop dramatically.

Likewise, as indicated earlier, the cost of EVs themselves will continue to drop as companies gain experience in manufacturing EVs and benefit from economies of scale. But that takes time. As Stefan Juraschek, vice president of electric powertrain development for BMW, said in December 2016, "We simply have to walk through the valley of tears to figure out how to save more money on producing battery-powered cars."[26]

The question of when EVs will become cost competitive, taking into account the full cost of ownership—including future maintenance and energy savings—is difficult to answer. Bloomberg New Energy Finance analysts predicted in 2016 that the cost of owning smaller-sized battery electric cars would equal that of gasoline cars in 2022.[27] That prediction is more optimistic than most. International Council on Clean Transportation analysts, who uncovered the Volkswagen emissions cheating in 2015, forecast that EVs will be cost competitive in about 2025 for small EVs with less than 150 miles of range and in about 2030 for 200-mile-range cars.[28] The break-even point will be somewhat sooner in Europe, where fuel prices are much higher.

In any case, because batteries are expensive (and heavy), they will be most attractive in smaller vehicles with short driving ranges and least attractive in large vehicles. Fuel cells and hydrogen require more space than gasoline and diesel engine systems, but much less space than batteries. Thus fuel cells are likely to prove more attractive than batteries in larger vehicles, as shown in figure 2.6.

A landmark study at UC Davis attempted to quantify the extra costs to the economy of transitioning to EVs (including plug-in hybrid and fuel cell vehicles). The researchers conducted a very detailed analysis of energy and vehicle costs that included price projections for petroleum, electricity, and hydrogen. They accounted for scale economies and forecasted costs of electric charging and hydrogen fueling infrastructure. They found that total transition costs for the United States would be $300 to $600 billion over a twenty-year period.[29] Although this sounds like a lot of money, the authors note that "when these estimated transitional investment and subsidy costs are compared to the base cost that

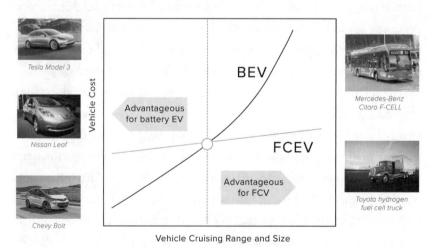

Figure 2.6. Relative advantages and costs of battery versus fuel cell EVs.

all U.S. consumers spend on new vehicles and fuels for light-duty vehicles (about $1 trillion per year, or $20 trillion over 20 years), the cost is modest, even small"—roughly 2 to 3 percent. They note that the benefits from energy savings, not including greenhouse gas reductions, are likely to far outweigh the extra capital costs well before the twenty-year transition is completed.[30]

Pushback from Vested Interests

Besides being slowed by consumer caution and high costs, the transition to EVs will be gradual for another real-world reason: there will be those who attempt to undermine public support and consumer interest for ideological or financial reasons. They will use a variety of means to slow government support and dissuade consumers. Especially challenging is the oil industry. In a very fundamental and direct way, EVs destroy the business of oil companies. An article in *Alberta Oil* magazine in July 2015 entitled "Is Tesla's Model S the Beginning of the End for Oil?" asserted, "Creative disruption has already wracked most major industries, and it's wrecked more than a few of them in the process. If it's going to visit itself upon the fossil fuel industry, it's almost certain to take the form of an electric vehicle."[31]

Some large oil and gas companies are exploring ways to engage with EVs, but for the most part, they see little that is appealing. They could sell natural gas to electric utilities, which they increasingly are doing. But they don't want to get into the electricity business, which they see as a low-margin, heavily regulated industry. They could get into the battery business, but that does not fit their core competencies. They are essentially highly capital-intensive companies that assemble massive amounts of capital to construct massive facilities—pipelines, refineries, and offshore oil drilling platforms.

One EV opportunity that might prove attractive to oil companies is the production of hydrogen from fossil energy. Oil companies could convert natural gas or coal into hydrogen and capture and sequester the carbon that remains, creating a near-zero-emissions energy source. The economics could prove attractive as governments start to impose taxes and caps on carbon and as more experience is gained with carbon capture and sequestration. So far, though, the oil industry has been slower to invest in low-carbon technology than other key industries—including both automakers and electric utilities.

In any case, the erosion of the oil business will be gradual. EVs are most attractive in the light-duty vehicle market, which accounts for less than half the total oil consumption in the world. The larger the vehicle, the more challenging (and slow) the embrace of electrification will be. Thus the changeover to electricity and hydrogen by large trucks will be much slower, and slower still for trains and ships. And still remaining for the oil industry is the large petrochemical market, where oil and natural gas are converted into chemicals and plastics for a wide variety of industrial and household applications. Biofuels will undoubtedly replace oil eventually for many trucks, planes, ships, and chemicals, but many issues confront biofuels as well, including cost, availability, and the large amounts of land needed.

Also slowing the demise of the oil business is the likelihood that fossil fuel subsidies—as much as $5.3 trillion annually worldwide, according to the International Monetary Fund[32]—will not shrink much or fast. Renewable energy subsidies, in comparison, total only about $120 billion each year.

What Can Be Done to Speed Consumer Embrace of Electric Vehicles?

The situation as it now stands is that automakers have made major investments in EV technology, but EVs are still significantly more expensive than comparable gasoline vehicles. Not surprisingly, most automakers are reluctant to scale up—to make a full commitment to EVs. Automakers are looking to consumers and government policy to further embrace EVs.

In contrast with the shared mobility and automation revolutions, which are being pulled by consumer demand (and industry competition), vehicle electrification is being pushed by policy (even Tesla is heavily dependent on government regulations and subsidies). Policy makers around the world embrace EVs to the extent that they view them as being in the larger public interest, solving local air pollution and global climate change, reducing oil imports, and/or (especially for China) leapfrogging the legacy international automotive companies. They are adopting a raft of incentives and regulations to push EVs into the market. But which policies and regulations should they be adopting?

Tax credits and rebates can directly reduce the price paid by consumers, while fuel economy, greenhouse gas, and EV rules indirectly reduce EV prices by motivating automakers to discount the vehicle prices in order to sell them. Other incentives such as access to carpool lanes and public chargers provide further inducements to buy an EV. The research is still tentative on the economics, equity, and politics of these policies and the urgency of enacting them; however, some insights are clear.

Perhaps the most obvious and effective policy is direct payments to consumers.[33] Both China and Norway embraced this approach in the early years of the transition. China provided incentives as large as $30,000 per vehicle in some cities.[34] Norway provided even larger subsidies by waiving very high purchase taxes (amounting to as much as 100 percent of the car's price). For a Tesla in Norway, this comes to as much as $80,000. The United States provided substantial but more modest incentives, totaling as much as $13,000: a $7,500 federal tax credit plus rebates of thousands of dollars from many states and cities. Subsidies at this level are not sustainable as EV sales grow, even for affluent Norway.

A more sophisticated approach is the concept of "feebates," also known as "bonus-malus" in Europe. Fees are imposed on polluting, gas-guzzling vehicles, and rebates are offered to buyers of energy-efficient, low-carbon vehicles. Two attractions of this approach are that it is revenue neutral, meaning taxpayers are not asked to pay for incentives, and it is permanent: EVs and energy-efficient vehicles will always be advantaged, with no subsidies needed.

Jump-Starting the Electric Vehicle Market in Norway

Norway is home to the largest per-capita EV market in the world, with more than 118,000 battery EVs and about 49,600 PHEVs as of July 2017 in this nation of 5.2 million people.[35] A notable surge in EV adoption occurred between 2010 and 2017, with EVs accounting for nearly 35 percent of light-duty vehicle sales in 2017.[36] But this stunning growth didn't appear out of nowhere; it was the result of decades of cooperation among private enterprises, public authorities, and NGOs. Driving forces were a Norwegian parliament united in its resolve to reduce greenhouse gas emissions and meet its UN climate goals, a couple of homegrown Norwegian EV manufacturers (Think Global and Pure Mobility) that seeded the market, and an abundant supply of clean hydropower.

Norway's project to wean drivers off fossil fuels is notable not only for its scale but also for the fact that the country is one of the world's largest producers of oil and natural gas. The country has strategically invested its fossil fuel wealth to forward-fund the future by providing significant financial incentives to EV owners. Christina Bu, the head of the Norwegian Electric Vehicle Association, says the country's system works because "it's constructed to make the least-polluting cars the most attractive"[37] while making it financially painful to keep driving anything that runs on gas or diesel.

Starting in the early 1990s, a broad coalition of different political parties has gradually introduced a substantial package of incentives to promote zero-emission cars in Norway.[38] The incentives include

- exemption from purchase/import taxes that are among the highest in the world (1990),
- low annual road tax (1996),
- exemption from charges on toll roads or ferries (1997 and 2009),
- free municipal parking (1999),
- access to bus lanes (2005),
- 50 percent reduction in company car tax (2000),

- exemption from the 25-percent value-added tax (VAT) on purchase (2001), and
- exemption from the 25-percent VAT on leasing (2015).

In 2015, the political parties agreed to maintain financial incentives until 2018 and then begin transitioning to a regime in which all vehicles are taxed based on the external costs they impose on society. EV owners will start paying half of yearly road fees in 2018, increasing to the full rate by 2020, and local authorities will decide whether EVs can park for free and use public transport lanes. Free toll roads will probably be replaced with differentiated prices depending on a vehicle's greenhouse gas emissions. Meanwhile, charging infrastructure will continue to be built as a result of a program launched by the government in 2009.

The Norwegian parliament adopted a goal that all new cars sold by 2025 should be electric or hydrogen vehicles or plug-in hybrids. To achieve this goal, the government plans to strengthen its green tax system based on the principle that the polluter pays.[39] Christina Bu summarizes, "Our role as a beacon in the international electric vehicle market is perhaps the most important climate initiative Norway contributes in the long term."

Nonmonetary incentives can also be highly effective and have the advantage of not imposing direct burdens on taxpayers. One particularly attractive nonmonetary incentive—valued in some studies at $5,000 or more[40]—is allowing EVs to use carpool lanes, even with a single passenger. But carpool lanes are not available everywhere and, in any case, would get quickly clogged. In California as of 2016, 270,000 registered EVs out of nearly 25 million registered passenger vehicles in the state were awarded this privilege. An even more valuable nonmonetary incentive is the ability to avoid lotteries and auctions for new vehicle registrations where they exist, as in major cities of China. Given that the auction price in Shanghai was around $15,000 in 2017 and that less than 1 percent of bids for new cars were being accepted in Beijing, the value of this Chinese incentive is huge. For most, it is the difference between owning a car and not.

Another effective subsidy is the installation of public electricity chargers and hydrogen stations in the early years of the transition. These are critical for two related reasons: (1) the psychological value of a large, visible network of charging stations is huge in convincing prospective buyers that they will not be stranded; and (2) the revenue from electricity sales is so small, especially for electricity chargers, that there is no compelling business model for investment in public chargers. In practice, most people charge mostly at home and thus make little use of public chargers, further eroding the business model for public chargers. Home charging is not only convenient but also often subsidized, as in the United States, where federal and state tax incentives are available for purchasing and installing 240-volt home charging systems.[41]

Perhaps the easiest path toward away-from-home chargers is for employers and retail establishments (such as shopping malls and hotels) to subsidize chargers (and the companies that install and operate them) as a fringe benefit for their workers and customers. As the EV market expands to residents of apartments and condominiums, the use of public chargers will increase. The most compelling public chargers are high-power versions that will charge a vehicle in thirty minutes or less, versus four to eight hours for 240-volt chargers. These fast chargers are far more expensive and require more safety precautions. But they are most urgently needed in order to convince buyers that they can recharge any time and use their vehicles for long trips. They are also needed to support the intensive use of EVs by shared mobility companies such as Uber and Lyft.

With charging and fueling, the main questions are how many stations, when and where, and who pays. In August 2016, Nissan released a report arguing that based on current trends in the United Kingdom, EV charging stations there would outnumber gasoline stations within four years.[42] In late 2016, Ford, VW Group, BMW, and Daimler joined forces to set up a network of fast-charging stations for EVs in Europe.[43] In Japan, where the greatest commitment has been made to public charging, more than six thousand fast chargers were in place by 2016. They were funded through various public-private arrangements.[44]

The case for hydrogen stations and the need for initial subsidies is similar. While the stations are substantially more expensive than EV chargers (more than a million dollars each), the difference is that eventually the hydrogen stations will become profitable because the refueling time (and thus customer turnover) is quick and the revenue per fill is much greater. Plus, drivers will not have the option of fueling at home.

But again, who pays the initial subsidy for hydrogen stations? There are roughly 150,000 gasoline stations in the United States, but as of 2017, fewer than forty hydrogen fueling stations, almost all in California. The State of California has committed $20 million per year through 2022, including some leveraged funds from Toyota and Honda, which will result in close to a hundred stations. But much more public subsidy is needed because hydrogen stations will not be profitable for many more years, and oil companies—with the notable exception of Shell—do not see a significant competitive advantage to being a pioneer (unlike automakers with EVs).

Another effective intervention by government to support EV sales is special treatment of EVs in the vehicle regulatory process. The United States, Japan, China, and the European Union, among others, impose aggressive fuel economy and greenhouse gas performance standards on new cars and sometimes trucks. All provide special provisions that make EV sales attractive to automakers. For instance, the United States and Europe allow EVs to be rated as zero grams per mile of emissions and even to count double in determining compliance with the greenhouse gas regulations (though that incentive is scheduled to be phased out in 2022).

Government leaders can also encourage consumers to embrace EVs using their bully pulpit. This is a way to provide confidence to consumers and industry that government really is committed to nurturing the transition to EVs. Reducing risk and uncertainty—for consumers as well as industry—is hugely important in the early years of a transition.

How exactly the transition to battery, plug-in hybrid, and fuel cell EVs will unfold is unknowable. It will vary greatly among regions. Some strategies and paths are more expensive and complicated than others. But generally, the best policy approach is to provide visible incentives to consumers initially, ensure that a network of public chargers and hydrogen stations is prominently accessible, offer incentives to automakers, mandate sales where politically acceptable, and be adaptive and agile as consumer preferences, technological innovations, and industry investments evolve.

Beyond Electrification

EVs of all types are highly attractive for addressing air pollution and climate change. In general, battery electrification is best suited to cars, plus buses and trucks used for short routes, especially in polluted areas. It is most attractive environmentally where the electricity grid is most decarbonized. Hydrogen fuel cell electrification is generally best suited to bigger vehicles, such as large SUVs, pickup trucks, and vans. Plug-in hybrids could serve all these applications as well.

While all three types of EVs are likely to flourish in different ways in different places and at different times, by themselves they are not sufficient to create sustainable and carbon-free transportation. One important concurrent strategy is to decarbonize electricity and hydrogen production. But it will be a very long time before the production and transport of electricity and hydrogen truly approach zero emissions.

A second strategy is to reduce overall use of vehicles. Most cities cannot accommodate more vehicles, whether they are electric or not. It has become obvious from Los Angeles to Paris to Beijing to Sao Paulo that high dependence on cars, even if they are zero-emission cars, is neither sustainable nor desirable. One of the best ways to reduce vehicle use is by pooling and sharing rides, particularly in automated vehicles (AVs). Thus the goal highlighted in this book is to integrate electrification, sharing, and automation of vehicles.

Key Policy Strategies

The aim is to move both consumers and automakers toward a broader and more rapid acceptance of EVs.

At the National Level

- Impose feebates—rebates for buyers of EVs and fees for buyers of petroleum-powered vehicles—to provide permanent incentives for EV purchases and compensate for low gasoline prices that result in part from more EVs and less demand for oil. Make the feebates revenue neutral so that there is no cost to taxpayers.
- Offer federal incentives and tax breaks to employers and retail establishments to install chargers.
- Fund education and outreach to increase awareness of EVs and their benefits.
- Require government fleets to convert an increasing portion of their vehicles to EVs as vehicles are replaced.
- Impose a ZEV mandate on automakers, requiring them to sell an increasing number of EVs (including fuel cell vehicles) until sales account for about 15 to 20 percent of vehicle sales, but include some flexibilities such as credit trading.
- In fuel economy and greenhouse gas emissions regulations, assign a rating of zero emissions to EVs and fuel cell vehicles until they reach 15 to 20 percent of total vehicle sales. After that time, include upstream emissions in the rating.

At the State or Local Level

- Eliminate registration fees for EVs for low- and middle-income households until EVs exceed some percentage of vehicle sales.
- Give EVs free access to high-occupancy-vehicle (HOV) and toll lanes (initially).
- Subsidize public charging with a focus on fast chargers.
- Subsidize the construction and operation of early hydrogen stations.
- Subsidize the purchase and installation of home chargers, with a special focus on multifamily dwellings.
- Partner with utilities and public utility commissions to provide special rates and smart chargers that enable EVs to provide extra storage for the

electricity grid (and thus reduce the need to build more capacity to meet peak demand).

- Encourage employers and retail establishments to install chargers by offering state and local incentives and tax breaks.
- Require government fleets to convert to EVs as vehicles are replaced.
- Incentivize drivers for ridehailing companies to use electric vehicles.

CHAPTER 3

Shared Mobility: The Potential of Ridehailing and Pooling

Susan Shaheen

Shared mobility with pooled rides is the key to leveraging vehicle electrification and automation to reduce congestion and emissions and create livable urban communities.

R IDESHARING IS OLDER THAN HORSE-AND-BUGGY TRAVEL, AND recent innovations make sharing easier, more convenient, and more efficient than ever before. Innovative mobility services premised on pooling can lower travel costs, mitigate congestion, and reduce greenhouse gas emissions. They also offer travelers more mobility choices beyond the traditional bookends of auto ownership and public transit. While the realm of shared mobility is vast, including shared bikes, scooters, and cars, the focus of this chapter is on pooled services—placing more people in a single vehicle. Doing so unlocks huge economic, social, and environmental benefits.

The motivation for pooling is simple. First and foremost is econom-ics. Cars are among the most underused capital assets in our economy, sitting empty 95 percent of the time and carrying only one individual much of the remaining time. If a car were used more than 5 percent of the time, and if that car carried two, three, or four passengers, the cost per rider would drop dramatically. The benefits go well beyond cheaper mobility. Because the car would be carrying multiple riders who might otherwise be driving, there would also be fewer vehicles on the road, less parking space required, less air pollution, and reduced energy use and greenhouse gas emissions. Given that the world has more than one bil-lion cars and light trucks, the potential for major reductions in pollution and greenhouse gases is substantial—in the United States and elsewhere.

The transition to a future where many rides are shared is now possible. What remains to be seen is whether and under what conditions people will be willing to embrace sharing and pooling.

Historic Trends About to Be Disrupted

Shared mobility is a radical departure from the auto-ownership culture that became entrenched after World War II when manufacturing shifted from defense equipment to consumer goods. This culture spanned the industrialized world and was supported by the growth of high-capacity, high-speed highways and auto-oriented service industries, such as drive-through restaurants and drive-in movie theaters. Almost everywhere, car ownership increased and public transit use declined, often despite efforts to boost transit ridership.

Since the late 1960s, public agencies, particularly in the United States and Canada, have tried to increase the use of carpooling and vanpool-ing. These strategies have included trip-reduction ordinances, construc-tion of carpool lanes and park-and-ride lots, and the use of telephonic and computerized ridematching. How well have they succeeded? In the United States, carpooling peaked during the energy crisis of the 1970s.

The share of Americans carpooling to work was 20 percent in 1970[1] and then dropped steadily to only 9 percent in 2014[2] (see figure 3.1).

But new ways of facilitating shared rides, eventually aided by vehicle automation, might reverse historic trends. Smartphones, wearable technology, location-based services, social networking, and the mobile Internet are key enablers that make it easier for travelers to share rides. In the words of Lyft president and cofounder John Zimmer, "A full shift to 'Transportation as a Service' is finally possible, because for the first time in human history, we have the tools to create a perfectly efficient transportation network."[3]

Innovations in pooling are rapidly gaining market share, pushed by consumer demand. Around the world, on-demand transportation is booming. Even automakers are beginning to realize that cars will be used differently in the future. Companies such as Ford, GM, Mercedes, Fiat Chrysler, BMW, PSA, and Volvo are openly discussing the need to evolve beyond manufacturing driver-centric cars to become mobility companies.[4] Google (Waymo), Tesla, and Ford have all said that their plan is to put the first fully driverless vehicles into fleets for shared

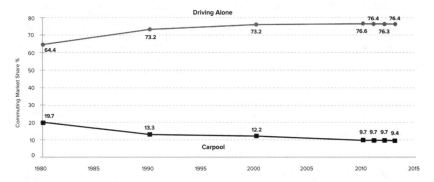

Figure 3.1. The decline in carpooling and the growth in commuters driving alone in the United States, 1980–2013. Source: Steven Polzin and Alan E. Pisarski, "Brief 10. Commuting Mode Choice," *Commuting in America 2013* (Washington, DC: AASHTO, October 2013), figure 10-2.

services. Ford's then CEO Mark Fields, pushing back against assertions that private vehicle ownership is the very definition of freedom, asked at the 2016 Consumer Electronics Show, "But is sitting in traffic in LA really freedom?"[5]

This begs the question, Which provides more freedom, driving a private automobile or being chauffeured? Posing questions such as this is how fundamental change starts.

The Growth of App-Based Pooling

Ridehailing services using smartphones got their start in 2010. UberCab (founded by Travis Kalanick and Garrett Camp and soon to become Uber) rolled out an iPhone app that provided a convenient way to hail a taxi in San Francisco. The app let travelers summon a black limo with the tap of a finger and pay for their ride seamlessly with a credit card. The service soon expanded to other cities and countries. In February 2013, Uber introduced UberX, a more affordable service using personal vehicles driven by their owners—a peer-to-peer-based platform. The company followed up with UberXL, offering larger-capacity SUVs or minivans that carry up to six people, and UberSelect, which uses luxury cars.

Another market player, Lyft (formerly Zimride), has a slightly longer history. Its precursor was founded in 2007 by Logan Green and John Zimmer, who were motivated as much by social and environmental aims as business goals. Zimmer has written that "ridesharing is just the first phase of the movement to end car ownership and reclaim our cities."[6] He points out that we have built our cities entirely around cars, and if we found a way to take most cars off the road, we would have a world with less traffic, less pollution, and more real community—a world built around people. Green's and Zimmer's initial company, Zimride, focused on matching riders for long-distance carpool trips. The service made the vast majority of its money from ridematching software that was run privately through individual schools and companies. Then in mid-2012,

Green and Zimmer launched their Lyft ridehailing app (months before Uber unveiled its similar UberX service)—described as a way to "unlock unused cars, create economic opportunities, and reduce the cost of transportation." By 2013, more than 130 university and corporate campuses were offering ridesharing and carpooling through Zimride, but the Lyft part of the business was growing much faster. In mid-2013, Green and Zimmer sold Zimride to Enterprise Rent-A-Car so that they could focus on Lyft.

Although Lyft and Uber, both headquartered in San Francisco, originally had very different values and approaches, they converged at about the same time on the same business model—taking a transaction fee from individuals driving their own vehicles. They were not alone in this space. They were accompanied by San Francisco–based Sidecar, which also launched operations in mid-2012. Ultimately, Sidecar left the market, selling its assets to GM in 2015.

Lyft, Uber, and Sidecar had few capital assets. They were mostly agile computer companies that did not need to manage large inventories of equipment, facilities, or employees. Their principal innovation was using computer algorithms that matched riders and drivers efficiently. Their apps removed the exchange of money from the rider-driver relationship—with fares automatically calculated and billed through the apps—and applied basic principles of economics to balance supply and demand by raising prices when demand exceeded supply (called surge pricing by Uber). The companies established a system for passengers to rate drivers and for drivers to rate passengers, with the driver ratings displayed prominently to riders before they accept the ride.

Lyft and Uber were soon joined by other mobility companies, such as Ola Cabs in India, Grab in Southeast Asia, Chauffeur Privé in France, and Didi Chuxing in China (which bought Uber's China subsidiary in 2016 and soon became the largest on-demand company in the world). All embrace the asset-light, peer-to-peer model of individually owned cars.

Also proliferating are specialized providers, such as Lift Hero (rides for older adults and those with disabilities) and HopSkipDrive and Kango (rides for children to and from school). These providers serve specific market segments (youth, older adults, passengers with disabilities) by using specially trained drivers who can assist riders with special needs and specialized vehicles that offer child seats and wheelchair accessibility.

Among the most transformative services are those that involve pooling—getting multiple riders into the same vehicle. There has been much experimentation. Some pooling services have been more successful than others.

Uber launched UberHop—an on-demand service for peak-period travel whereby travelers request their ride through the app, walk to a designated pickup location, and share the ride with other commuters—in cities around the world. Lyft piloted its Lyft Carpool service, whereby commuters pick up strangers along their route, in the San Francisco area. UberHop was eventually discontinued in all markets except for Manila, and Lyft Carpool was discontinued in less than a year due to low match rates.

More successful have been Lyft Line and UberPool, both launched in 2014. These services bring together previously unacquainted riders with similar origins and destinations. Computer algorithms assign additional riders to drivers in real time. The computer optimizes route changes as new pickups are requested, aiming to minimize the detours experienced by each rider. In return for the slight delay in getting to their destination, riders typically get a 40 to 50 percent discount on their fare, even if the driver never picks up another rider.

One variation of Lyft Line and UberPool is Driver Destination, offered first by Lyft and then Uber. In this mode, drivers enter a destination into the app and receive ride requests from along their commute. Another pooling innovation by those companies was to encourage passengers to walk a short distance to a major arterial street or to meet other riders at select intersections "hot spots"—in return for a discounted fare. Still another pooling innovation is matching of commuters traveling along major highways—offered in Washington, DC,[7] Chicago,[8] Chengdu in China, New Delhi and Bengaluru in India,[9] and probably elsewhere. While the Lyft Line and UberPool services continue to expand at a rapid rate, the variations described here have had mixed success—at least so far. Ridehailing companies continue to experiment as they seek ways to build pooling services.

Pooling can also be successful for longer intercity trips, as demonstrated by BlaBlaCar, the world's largest long-distance ridesharing service.[10] BlaBlaCar was founded in France in 2006 as a free platform for carpooling but transitioned in 2011 to a fee-based service. It connects drivers and passengers willing to travel together between cities and share the cost of the journey. Passengers and drivers are connected through a website that combines social media and a reservations platform in a way that enables a feeling of trust and safety, which is key to the company's growth. By 2017, BlaBlaCar had more than forty million members across twenty-two countries.[11]

App-based pooling is at an early experimental stage. The question is how to leverage these technologies to carry more passengers in more vehicles so that companies and society both win—companies generate more profit, and society benefits from reduced emissions and traffic congestion. The answer is in the details. Executives at Lyft and Uber believe that the key to their long-term success is lower prices and that the best way to reduce prices and costs, for now, is with pooling—putting more riders into a single vehicle.

Public support for pooling can take many forms—including, for instance, lowering travel times for pooled riders by giving priority access to pooled vehicles at curbs. Also needed is more experimentation with different types of services described earlier, along with public policies that encourage pooling, as highlighted at the end of this chapter.

Pooling in the San Francisco Bay Area

In the San Francisco Bay Area, commuters often use casual carpooling to get from the East Bay to downtown San Francisco during the morning commute. Casual carpooling, also known as slugging, is an informal system where people line up in self-organized locations to catch a ride with drivers wanting to use a carpool lane or avoid bridge tolls. Slugging is especially attractive to many Bay Area commuters crossing the San Francisco Bay Bridge because cars and passengers can access carpool lanes and enjoy reduced travel times, shorter wait times at toll plazas, and reduced tolls. Some say that time and cost savings are the "secret sauce" of casual carpooling—in this case, with average wait times for riders of just two and a half minutes.[12] Casual carpooling got started in 1979 due to a ninety-day Bay Area Rapid Transit (BART) strike and has grown organically since. It carries more than six thousand people per day.[13]

The Bay Area is also the home of Lyft and Uber, and it served as an initial test bed for their pooling services. A 2014 survey of 380 ridehailing customers in San Francisco found that half of the reported ridehailing trips had more than one passenger (not including the driver), with an average occupancy of 2.1 passengers (in contrast to 1.1 in taxis). Close to 90 percent of respondents said they waited ten minutes or less for a vehicle to arrive after requesting one, and 67 percent waited five minutes or less.[14]

With conventional (nonapp) carpooling stagnant[15] and congestion on the Bay Area's freeways and public transit approaching near-record highs,[16] the Metropolitan Transportation Commission (MTC), the region's transportation agency, believes that filling empty seats in private vehicles might be the most cost-effective way to increase capacity.[17] Between May and November 2015, MTC issued two calls to partner with private-sector app-based pooling providers. MTC executed pilot partnership agreements with four private-sector app-based services: Carma, Lyft, MuV, and Scoop. These services are testing marketing and incentive programs that encourage commuters to use new app-based services to increase pooling.

If MTC determines this pilot program is successful, the agency might decide to phase out its public ridematching services. MTC estimates it could save $500,000 annually by transferring ridematching services to the private sector. MTC is considering a number of other future innovations, including integrating app-based matching services with call-in transportation services, establishing designated "hot spots" for casual carpooling, integrating app-based services with park-and-ride facilities, and leveraging pooled services to access BART and other line-haul transit services.[18]

Why Is Pooling So Important?

When Uber unveiled UberPool, the company called it "a bold social experiment" and promised to get it right, "because the larger social implications of reducing the number of cars on the road, congestion in cities, pollution, parking challenges . . . are truly inspiring."[19] In a speech in London on UberPool's launch, then CEO Travis Kalanick said his goal was to take a third of that city's three million cars off the streets, which he said would not only reduce congestion and the capital's carbon footprint but also create one hundred thousand new jobs and dramatically expand the local economy.[20]

A study conducted by the Paris-based International Transport Forum in 2016 offers a glimpse into how shared mobility might change urban living in the future. This study modeled the impact of replacing all car and bus trips in Lisbon, Portugal, a midsized European city, with mobility provided by fleets of shared automated taxis and shuttle buses.[21] Among the key findings were that 97 percent fewer vehicles (cars, shuttle buses, and full-size buses) would be needed to serve all trips, 95 percent less space would be required for public parking, and 37 percent fewer kilometers would be traveled by the vehicles. The study also estimated that each vehicle would travel ten times the total distance traveled by current vehicles. The benefits of intensive use of vehicles are large: much lower cost per passenger (since depreciation and operating costs are spread out over many more occupants) and more rapid turnover of vehicles, which results in accelerated adoption of cleaner vehicles.

A study of New York City by researchers at the Massachusetts Institute of Technology came to similar conclusions: three thousand four-passenger cars could serve 98 percent of taxi demand in New York City, with an average wait time of only 2.7 minutes.[22] The researchers also found that 95 percent of taxi demand would be met by just two thousand ten-person vehicles, compared to the nearly fourteen thousand taxis that currently operate in New York City. They used an algorithm that reroutes cars in real time to serve incoming requests.

Finally, a study by researchers at the Lawrence Berkeley National Laboratory found that a fleet of shared automated electric vehicles, with rightsizing of vehicles by trip—that is, smaller vehicles for fewer passengers and larger for more passengers—in combination with a low-carbon electricity grid (forecasted for 2030), could reduce per-mile greenhouse gas emissions by 63 to 82 percent by 2030 compared to privately owned hybrid vehicles.[23]

While these are extreme scenarios, they do indicate the potentially huge transportation, infrastructure, environmental, and social benefits of on-demand pooling. These future scenarios imply a shift away from personal vehicles, a transition that can be facilitated by one-way and peer-to-peer carsharing, whereby people borrow a car from a fleet of commercially or personally owned vehicles and drive themselves. Carsharing services represent a critical step toward creating more choice for travelers and making it easier for drivers to give up personal car ownership. But carsharing typically involves one customer (the driver) per trip in cars that are shared sequentially. Pooled rides have a far smaller carbon footprint, consume much less road and parking space, and have the potential to serve far more trips. In short, pooling is critical to maximizing the benefits of shared mobility.

This point is brought home by an analysis of the app-based ride services in New York City. The analysis found that despite heavy promotion of pooled options, single-passenger trips still predominate, and these services have added significantly to vehicle miles traveled on city streets.[24] The report, released in February 2017, stated, "After accounting for declines in yellow cab, black car, and car service ridership, [app-based ride services] have generated net increases of 31 million trips and 52 million passengers since 2013," and this has increased vehicle travel on the city's roadway network by six hundred million miles. This is clearly not a sustainable trajectory. More effort is needed to encourage pooling.

The Evolution of Other Shared Modes

Existing services, like taxis and public transit, naturally feel threatened when new competition emerges. Taxis are more vulnerable than public transit, which has many opportunities to partner with and benefit from app-based services. How are these services evolving to compete?

Taxis

The traditional taxi industry is clearly threatened by ridehailing services. As just one example, the number of taxi rides in San Francisco, where Uber and Lyft got started, fell by 65 percent between March 2012 and July 2014, and in January 2016, the city's largest taxi company, Yellow Cab, filed for bankruptcy.[25] From New York to Paris, taxis have been fighting to block Lyft and Uber, with limited success. Governmental protection helps taxis survive, but if they are to compete successfully with ridehailing, carsharing, and microtransit, taxis will need to upgrade their services.

Electronic hailing (e-hail) services, such as Arro, Bandwagon, Curb, Flywheel, Hailo, and iTaxi in the United States, are a step in that direction. Travelers can use these mobile apps—maintained by either the taxi company or a third-party provider—to digitally dispatch a taxi. These services are gaining popularity. As of October 2014, Flywheel was used by 80 percent of San Francisco taxis. In April 2017, Flywheel was acquired by taxi telematics provider Cabconnect, which hoped the deal would create a universal app for booking taxis.[26] By May 2017, Curb (acquired by Verifone) was serving sixty-five US cities with fifty thousand cabs.[27] Increasingly, taxi and limousine regulatory agencies are developing e-hail pilot programs and mandating e-hail app compatibility.

This e-hail innovation, although in the works for many years, has been embraced largely in response to the success of Uber and Lyft. Where this innovation has been adopted, taxis have reduced their wait times to closely match those of their shared-ride competitors. But e-hail services still face market challenges, including matching supply and demand and the requirement to charge locally regulated taxi rates. Furthermore, taxis cannot employ surge pricing during periods of high demand, as ridehailing companies often do.

Another innovation is taxi sharing. Multiple passengers are desirable not only to utilize taxis better but also to reduce fares so that more people can gain more access and mobility. Until now, though, few taxis were allowed to carry multiple unrelated passengers with different origins and destinations. Cities like Los Angeles, Burbank, and Boston have permitted taxi sharing only in downtowns and at airports. New York City technically allows taxi sharing, but in practice, it has gained acceptance only at airports, at some in-city taxi stands, and along one eastside corridor.[28]

Shared-ride taxis arose in different places for different reasons. In Washington, DC, they were initiated during World War II to conserve scarce resources. In New York City, the Share-a-Cab service was started at LaGuardia Airport in 1979 to address the problem of long lines at the cabstands. A combination of an open marketplace, aggressive marketing, economies of scale, and good-quality, reliable service resulted in cost-effective shared taxi services in places like Little Rock, Arkansas, where in 1977, shared-ride taxis were transporting more than 1.7 million passengers annually.[29] But since then, local regulations have curtailed and often eliminated these services.

In July 2013, Bandwagon reintroduced shared-ride taxis in New York City. In a partnership with the New York City Taxi and Limousine Commission, Bandwagon provides an iPhone app that lets riders share taxis at LaGuardia Airport, JFK Airport, Newark Liberty Airport, New York Penn Station, and the Port Authority Bus Terminal. Waiting passengers text Bandwagon their destinations and are paired with passengers with similar routes and destinations. Paired passengers are permitted to advance to the front of a taxi line, get into their cab, and split the fare. Bandwagon claims that the app contributes to shorter taxi lines, reduced wait times (when a user at the end of the line is paired with a passenger at the front of the line), and cost savings of up to 40 percent per taxi trip.[30]

More evidence for the potential benefits of taxi sharing comes from a 2010 experiment in New York City in which taxis transported up to four passengers at a reduced fare of three to four dollars.[31] The shared taxis picked up passengers at designated pickup locations and allowed passengers to get off anywhere along the route during the morning commute. The experiment was credited with making taxi sharing more convenient, increasing taxi capacity during peak commute periods, providing cost savings to passengers, and lowering greenhouse gas emissions.[32] A subsequent study in 2014 quantified the impacts of taxi sharing in New York City and found that it could reduce taxi trips by an estimated 40 percent and thereby cut the associated carbon dioxide emissions.[33]

Public Transit, Microtransit, and Paratransit
Public transit ridership has declined in most US cities due to a number of factors, including low fuel prices, poor transit performance in some markets, and, in some cases, competition with ridehailing services.[34] Public transit operators are under tremendous pressure to improve the quality and quantity of service as more cities focus on social equity, urban livability, air quality, climate change, and congestion. Partnering with shared mobility operators is one strategy to achieve these goals. A combination of nonprofit and for-profit services can provide pooled bus-like services and facilitate first- and last-mile connections, bridging gaps where the public transit network is sparse.[35] The evolution of these partnerships is examined in chapter 5.

Microtransit operators use vans and small buses to offer a broad mix of demand-responsive, curb-to-curb services. These services tend to be less expensive than taxis but more expensive than conventional bus and rail. Typically, riders use mobile apps to pay for trips electronically and track the vehicles as they approach, although a few microtransit services use telephone dispatch and cash payment mechanisms.

Microtransit has evolved from jitneys in cities like San Francisco and New York City. Jitneys are common in many cities around the world in various forms, but in the United States, they have largely disappeared as a result of rising insurance costs and regulations to protect public transit. They usually carry up to fifteen passengers in a van over a semifixed route, operating along busy streets and making numerous pickups and drop-offs. Some provide door-to-door service for an extra charge.

One jitney success is the "dollar vans" in New York City, which got their start in 1980 due to an eleven-day public transit strike. Dollar vans had a ridership of about 120,000 people per day in 2016.[36] They are a shadow transportation service that follows popular bus routes but also serves communities neglected by subways and buses. They often operate in low-income neighborhoods that have large immigrant populations. Since 1994, the city has been issuing permits to dollar vans in an attempt to regulate them, but most still operate illegally, without any formal regulation or oversight.[37] Because these services are well integrated into the community, regulators frequently condone them.[38] In March 2017, 325 official (licensed) dollar vans were running,[39] down from more than 1,000 just a few years earlier. The decline probably reflects a lack of license enforcement rather than an actual decline in number of vehicles.[40]

In recent years, new microtransit services have been emerging, offering varying levels of demand responsiveness on a mix of fixed and flexible routes.[41] Microtransit services in the United States include Chariot and Via. Chariot, which was acquired by Ford in September 2016, operates like a public bus service. It initially ran vans along predefined routes in Austin and San Francisco, with plans to expand to more cities. Customers can make requests for new "crowdsourced" routes that are created based on demand. Via, which operates in New York City, had approximately five hundred vehicles and provided more than 1.5 million rides from its system launch through October 2015. Bridj, a microtransit service that started in Boston in 2014, eventually extended limited service to Austin, Kansas City, and Washington, DC, but shut down in May 2017 as a result of low ridership and failing to strike a hoped-for deal with a major automaker.[42]

Microtransit has great potential, but considerable effort is required to make it successful. It might be particularly well suited to complement, enhance, or replace existing paratransit or dial-a-ride services, which are legislatively required to provide rides to passengers with mobility limitations. These services dispatch specially outfitted small buses and vans on request and operate door to door. Paratransit became common in the United States in the 1970s as requirements were imposed and subsidies provided to serve people with disabilities. Paratransit providers are typically embedded in larger transit bus operators or are small companies that contract with public transit operators, often outsourcing to taxis. Microtransit and paratransit are discussed further in the next chapter. The takeaway here is that these services are ripe for integration into a larger shared mobility system.

When Do People Choose Shared Rides?

Two big questions loom: When and at what price are people willing to share rides with strangers? And under what conditions are people willing to give up personal car ownership and replace those car trips with shared mobility services?

Early use of carpooling was motivated by the desire to conserve time, money, and/or fuel. In the United States, the government began to encourage carpooling during World War II as a means of reducing costs and gasoline use. The OPEC oil embargo of 1973–74, which led to cars waiting for gasoline in lines that snaked around the block, motivated many people to carpool. A survey of 197,000 employees during the energy crisis of the 1970s found that 15 percent of them became carpool commuters, resulting in a 23 percent reduction in vehicle miles traveled among survey respondents.[43]

Sharing and pooling serve an important role in enhancing mobility in low-income and immigrant households where cost, driving skills, or legal status might be an issue.[44] Data from the National Household Travel

Survey and the American Community Survey indicate that ridesharing users generally have lower incomes, and some minorities (typically Hispanics and African Americans) carpool more than other racial and ethnic groups. Both economics and culture influence people's willingness to share rides.

A 2016 study of Bay Area casual carpooling—where people line up in self-organized locations to catch a ride with drivers wanting to use a carpool lane or avoid bridge tolls—revealed that many people are willing to share rides with strangers in exchange for time and money savings and to reduce commute stress.[45] The study found that 75 percent of casual carpoolers were previously public transit riders, while approximately 10 percent previously drove alone.[46] An online survey of casual carpoolers in northern Virginia found similar motivations. The chief reasons for riding with strangers, according to this survey, were to save gasoline and to use time for other purposes during the drive, while the primary reason given for driving instead of riding was departure time flexibility.[47]

The evidence is strong, from surveys and anecdotal evidence, that consumers share rides when it is convenient and more affordable than driving alone. What is not known is how many people are willing to ride with strangers and under what conditions. Even less is known about how many people are willing to give up car ownership and under what conditions. Because survey research is notoriously unreliable in predicting future acceptance of unfamiliar services and products, it will take considerable real-world experience and research to accurately assess the chances for future success of new mobility services.

Adding Automation to the Picture

The attractions of shared mobility could be greatly expanded by integrating it with vehicle automation. Without a driver and with more intensive vehicle use, pooling could be much cheaper (costing perhaps 90 percent less than today's pooling services, as shown earlier in figure 1.2). Many companies are looking forward to this future. Indeed, as part of a legal proceeding in April 2017, Uber said that any injunction that blocked it from continuing work on automated cars could harm its ability to be a viable business.[48] Uber has already bet hundreds of millions of dollars that autonomous cars are the future. CEO Travis Kalanick said, "If we weren't part of the autonomy thing? Then the future passes us by."[49]

By 2017, various small-scale shared automated vehicle (AV) pilots were emerging across the globe.[50] Uber's pilot in Pittsburgh and nuTonomy's in Singapore were joined by EasyMile's driverless shuttle on predefined routes in Finland, France, Switzerland, and San Ramon, California. City-Mobil2 launched automated demonstration projects in Italy and Finland and was planning to launch larger demonstration projects in France, Switzerland, and Greece. Local Motors in Tempe, Arizona, unveiled Olli, a self-driving electric microbus designed to work on college and corporate campuses and to fill gaps in urban transit systems (see figure 3.2).

Figure 3.2. Olli, the self-driving electric microbus by Local Motors. Photo courtesy of Local Motors.

Considering that fully automated vehicles might initially cost anywhere from $10,000 to $50,000 more than an equivalent nonautomated vehicle,[51] some analysts predict that the first AVs introduced to the public will be part of a shared-fleet service and not targeted at individual owners. Lyft, which received a $500 million investment from GM in January 2016, envisions a subscription model for a shared AV service.[52] In September 2016, Lyft cofounder John Zimmer boldly predicted that in five years, the majority of Lyft rides will take place in AVs, and by 2025, private car ownership will be rare in major US cities.[53] In October 2016, Tesla Motors described a future "Tesla Network" that will let its fully automated cars provide rides for a fee while the owner is not using the vehicle.[54] In Europe, Deutsche Bahn, the continent's largest railway company, based in Germany, plans to operate fleets of shared AVs that could be used for trips to and from their regional rail stations.[55]

Not surprisingly, US cities and public agencies have begun to examine possible ways to advance, test, manage, and regulate shared AV services. The federal government's Smart City Challenge sparked interest in emerging transportation technologies in cities across the United States, with seventy-eight cities completing the initial application for the $50 million award.[56] AVs were a key component of most proposals. Columbus, Ohio, won the challenge in June 2016 with a proposal that included a shared automated shuttle connecting public transit to retail districts. San Francisco proposed a plan for shared, electric, connected AVs to replace single-occupant-vehicle ownership and use.[57]

More research is needed to fully understand the opportunity for shared AVs. The mainstreaming of shared automated fleets will likely vary by region and be heavily influenced by local factors such as population density, urban form, local policies, and private vehicle ownership costs. What is clear, though, is that whatever is done to expedite and encourage pooling today will set the stage for AV pooling.

How Can We Create an Environment Conducive to Pooling?

Advancements in mobile computing and widespread use of smartphone apps provide new opportunities for pooling. For decades, public policy emphasized the construction of capital-intensive high-occupancy-vehicle (HOV) lanes and park-and-ride lots—and still, as we have seen, carpooling declined. While HOV capacity enhancements might provide the foundation for future pooling, digital infrastructure will likely be more important. Pooling that leverages modern information and communication technologies could increase vehicle occupancy and mitigate congestion on existing roadways, without the addition of HOV facilities.

Public policy will play an influential role in accelerating pooling in conventional, electric, and eventually automated vehicles. Cities and local governments can use their authority over land use and local infrastructure, including roads, rights-of-way, public parking, and curb space, to favor pooling. They own and subsidize public transit and, in many cases, regulate and tax the taxi industry; they can use those carrots to encourage partnerships and services that aid pooling.

Parking is illustrative. The current rise of shared mobility and the imminent rise of AVs reduces the need for parking at trip ends. Given this, local governments could consider revising or even eliminating minimum parking requirements. These parking requirements make expensive-to-provide parking free to drivers; they are large subsidies to personal vehicle ownership.[58] Eliminating these parking requirements would substantially reduce land development costs, and therefore housing costs, as well as the cost of nearly everything sold in commercial establishments. Reducing or removing parking requirements would also encourage travel by other modes, including shared AVs. Planners can work with communities to consider repurposing redundant parking and other road infrastructure as parks, trails, bike paths, and affordable infill housing.

One important challenge for pooling, especially in providing first- and last-mile connections to public transportation, is data sharing. Without access to travel data, local governments cannot effectively invest in transportation and parking infrastructure. They also cannot effectively manage traffic congestion, regulate competing services, design public transit routes, subsidize transit and paratransit, meet the needs of disadvantaged travelers, or plan future public transportation funding. Open data are also needed to package mobility service choices for use in apps. There are alternative ways of getting data, including traveler surveys and new techniques of collecting "big data" from mobile phones. But all those alternatives are expensive, slow, and/or unreliable.

To establish repositories for public- and private-sector data sets, the public sector needs to create firewalls that maintain confidentiality for mobility service providers and travelers. Government safety and environmental regulators do this routinely with automakers (who submit plans for future models and technology). The public sector can play a critical role in establishing and overseeing data standards, developing security protocols, and maintaining data exchanges. The US Department of Transportation's federal policy on AVs calls for sharing data generated by testing and deployment "in a way that allows government, industry, and the public to increase their learning and understanding as technology evolves but protects legitimate privacy and competitive interests."[59]

Moving forward, it is clear that people and cities are on the cusp of rapid change as a fundamental reimagining of transportation unfolds across the world. Pooling is central to this reimagining and to the creation of more economically, environmentally, and socially sustainable transportation.

Key Policy Strategies

The aim is to create a social and institutional environment that is conducive to pooling.

At the National Level

- In regulations for fuel economy, greenhouse gas emissions, and zero-emission vehicles (ZEVs), award extra credit to automakers for the sale of passenger-centric vehicles used for mobility services. These vehicles should be used principally for pooling services, with in-vehicle reporting or monitoring procedures for enforcement.
- Encourage mobility companies to modify apps to include safety-related profiles of passengers, which are made available to fellow passengers and include a rating scheme (similar to ratings of drivers and passengers with Uber and Lyft), with rules that prohibit discrimination based on race, ethnicity, gender, socioeconomic status, or disability.
- Give tax credits to mobility service providers for achieving average passenger occupancies of two or more for cars and higher for vans, with effective reporting and enforcement.

At the State Level

- Reduce vehicle registration fees for car owners and mobility service companies who use their vehicles for pooling (but not for single-passenger UberX-type services). Those that benefit from low or zero fees should agree to reporting or monitoring.
- Provide tax breaks for companies that demonstrate a high level of pooling among employees, with the threshold specified by location (for example, a higher threshold in dense city centers).
- Provide subsidies for low-income travelers using pooled services (including paratransit).
- Explore the use of standard open APIs (application programming interfaces) across shared-ride services to provide more information to riders.

At the Local Level

- Give pooled vehicles special parking and stopping privileges at airports and other congested areas.

- Provide discounts for pooled cars and vans on toll roads and toll lanes.
- Give pooled vehicles priority in urban areas, including designated curb space and shared mobility lanes, and allow free use of bus stops and taxi stands.
- Reduce onerous regulations on taxis so they can compete effectively, including allowing them to offer pooled services.
- Offer free carsharing coupons and public transit passes for workers who use pooling, along with rewards such as express lines in employer cafeterias.
- Require mobility providers (including those managing AVs) to give data to a publicly accountable, not-for-profit institution in a secure facility. Require that data to be made available in a time-delayed fashion in anonymized form to local and regional planning agencies and researchers, respecting competitive and personal confidentiality concerns.
- Ban new parking garages (with rare exceptions) and provide incentives to owners of private parking garages to allow only shared vehicles in private parking spaces at all business destinations. Offer assistance to these owners in transitioning private lots for redevelopment.
- Impose caps on parking space allowed for any new building, possibly even houses, reversing the practice of imposing minimum parking requirements on new developments (adopted in the past to keep cars from flooding existing curb spaces and parking garages).

CHAPTER 4

Vehicle Automation: Our Best Shot at a Transportation Do-Over?

Daniel Sperling, Ellen van der Meer, and Susan Pike

Automation could lead to dramatically safer, cleaner, more affordable, and more accessible mobility—but only if combined with electrification and shared rides.

FROM SAN FRANCISCO TO SHANGHAI, THE CAR of the future is automated. Automakers and tech companies are racing to develop, test, and bring cars to market with a dizzying array of new information and automation technologies. Every new model year brings more innovation and announcements about the future. But what exactly are these companies bringing to market, and how automated will these cars really be? Do we call them autonomous or automated or driverless, and what does that even mean? Will they really transform transportation as we know it? The spin is confusing and bewildering.

In this chapter, we untangle what is meant by "autonomous" vehicles, when these advanced cars are likely to be available, how they will be used, and what this revolution means for our lifestyles and cities—and pooling. We are humble. We can explain the different levels of automated technology, but no one—including us—knows precisely when the more advanced technologies will hit the road and even less when they will be widely adopted. We share what is known about how this new phenomenon is likely to unfold and our insights on how hype deviates from truth.

What we know for sure is that the technological arms race has begun. Companies are already competing to control the technology and market—with regulators and governments struggling to keep up.

What's the Buzz?
Unless you have been disconnected from all media, you know that automated vehicles (AVs) are coming. So much media attention has been lavished on this technology that expectations are sky-high and rising. Technology and automotive companies deserve much of the credit—or blame. While the history of AV technology spans many decades, autonomous cars did not capture the popular imagination until Google announced in 2010 that it was undertaking an experimental "self-driving car" project. Over the next few years, Google firmly implanted the phrase "autonomous vehicles" in the public vernacular. In 2013, Tesla inflated the hype, announcing the introduction of its Autopilot system (though the Autopilot software was not activated until 2015[1]). Nearly all automakers soon followed with an avalanche of announcements about when they would launch AVs. Other consultants and prognosticators piled on.

A 2013 report from Morgan Stanley said that sometime between 2018 and 2022, cars would have "complete autonomous capability" and that by 2026, autonomous vehicles would fully penetrate the market.[2] In 2014, IHS Automotive predicted that by 2050, nearly all vehicles on the road will be self-driving.[3] In 2015, Tesla announced that full autonomy was about three years away,[4] and Ford reported that it planned to have fully autonomous vehicles by 2021.[5] Similar announcements were made through 2017 by senior executives at GM, Volkswagen, Nissan, Google, Baidu, Intel, Lyft, Daimler, the Insurance Information Institute, the Institute of Electrical and Electronics Engineers, and more—but they were usually vague about what they meant by "self-driving," "automated," and "fully autonomous."[6] The hype was even coming from senior government officials. The former chief innovation officer at the US Department of Transportation, Chris Gerdes, said in 2016 that 35 percent of the cars on the road would be self-driving by 2026—which would require every single car sold from 2021 forward to be self-driving.[7]

Chris Urmson, technical lead at that time of Google's self-driving car program, told an audience at Austin's South by Southwest Conference in March 2016 that there was a little bit of truth in every one of these predictions, because the technology is going to be rolled out incrementally: "We imagine we are going to find places where the weather is good, where the roads are easy to drive—the technology might come there first. And then once we have confidence with that, we will move to more challenging locations."[8] This makes sense because of the many uncertainties— the rate of technological development, consumer adoption, legislative and regulatory changes, changes in physical infrastructure—but also because so many predictions lack specificity about the level of autonomy and driving modes.

Hype is not bad or wrong; it is a way to generate enthusiasm—among investors, politicians, future customers, and the media. It works. Hype about AVs has indeed stimulated innovation, bringing new investment and raising expectations. But it *is* hype. Gartner, a firm that monitors technology hype for its clients, posits that autonomous vehicles were just past the "peak of inflated expectations" in the second half of 2016 and wouldn't reach mainstream adoption for more than ten years (see figure 4.1).[9]

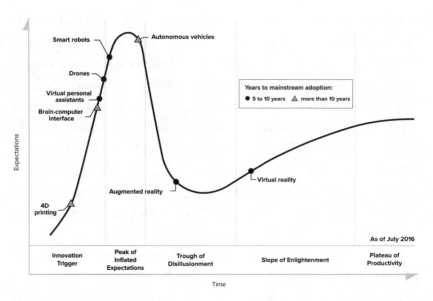

Figure 4.1. The hype cycle, with autonomous vehicle hype at its peak in 2016. Source: Gartner, July 2016.

At this point, we offer definitions of a few terms that are bandied around. *Autonomous* is a term that Google often used to promote its early bubble-like small cars (see figure 4.2). Reflecting the Silicon Valley culture of going it alone, Google emphasized that its vehicles would be able to operate without human intervention and without communicating with traffic signals, other vehicles, or other infrastructure. While that might be true in some situations, such as rural dirt roads, AVs will rarely operate in a truly autonomous fashion. They will likely be connected to other vehicles, to traffic signals, and to information beamed from roadsides. They will be automated and connected; thus we use *automated* as the generic descriptor in this book for a category that includes but is not limited to autonomous vehicles.

Other sloppy language includes *self-driving* and *driverless*. The distinction between the two is rarely acknowledged, but it is crucial. *Self-driving* means that vehicles do not need human input in some specific circumstances, such as on limited-access highways, where distractions

Figure 4.2. A Waymo self-driving car on the road in Mountain View, California, in February 2017. (Google's parent company, Alphabet, spun off its self-driving car unit as Waymo in late 2016.) Photo by Grendelkhan, CC BY-SA 4.0 (http://creativecommons.org/licenses/by-sa/4.0), via Wikimedia Commons.

and surprises are less common. Indeed, as highlighted later in this chapter, as of 2017, a few vehicles might be described as nearly self-driving, including the Tesla on Autopilot. They use adaptive cruise control to accelerate and decelerate, emergency braking to stop suddenly, and lane keeping to steer the vehicle within a lane. In some of these vehicles, if the human driver switches a turn signal on, the vehicle will switch lanes when it detects an opening. Soon, some vehicles will be able to decide on their own when to switch lanes. In these self-driving cars, the human in the driver's seat will be called on to retake control in certain circumstances—for instance, when a policeman directs a car to travel the wrong way on a one-way street to bypass an accident or construction. These cars will be capable of self-driving, but they will need to have a driver in the car for most or all of a trip. They will *not* be driverless.

A *driverless* car, by contrast, will not have a steering wheel, and the occupants will not need to have a driver's license. Indeed, the car will not need to have any occupants. It can be trusted to operate in all conditions and under all circumstances without human intervention. Driverless cars provide far more benefits than self-driving cars—but also potentially large downsides (including the specter of empty cars congesting traffic).

Why Are Vehicles Being Automated?

Are AVs inevitable and desirable? Inevitable, yes; desirable, probably. The most certain and clear-cut benefit of both self-driving and driverless vehicles is safety. Fully automated cars will be much safer than those with human drivers.

Safety has become a more compelling goal in recent years. Fatality rates had been dropping in the United States and almost all industrialized nations for more than a century. All that changed around 2012, just as smartphones were proliferating. Distracted driving is real. Because robots don't drink, eat, text, or sleep, distracted and drunken driving will disappear with automated cars. Google estimates that automation could eliminate at least half of the 1.2 million crash fatalities recorded each year around the globe.[10] One report predicts that accidents would drop by 80 percent by 2040.[11]

Other benefits, beyond safety, could also be large. Self-driving and driverless vehicles could greatly enhance access for young, old, ill, blind, poor, and disabled travelers, as well as provide less costly transport and more livable cities for everyone.[12] These benefits would be even greater if the vehicles are electrified and shared by multiple riders.[13]

There could also be economic benefits. Consider the time people waste each day driving their vehicles to work, school, errands, and social encounters. All this time could be freed up if people didn't need to focus on the task of operating the vehicle. For a typical worker, saving fifty minutes of commute time each day amounts to saving five forty-hour weeks each year. That saved time could be spent working—making the commuter more productive—or performing other activities. And even more time could be saved on shopping, social, and personal business trips. Travelers could work, watch movies, sleep, or just relax in the car. If we value people's time at $10 per hour in the United States, this would translate to several thousand dollars in savings per year per person.

Beyond time savings are potential infrastructure savings. Because AVs will drive more precisely, lanes could be narrowed, and because they will follow each other closely, fewer roads and lanes will be needed.

One of the greatest benefits of automation will be a reduction in parking space—if vehicles are operated by companies or nonprofit organizations and carry multiple riders. In Los Angeles County, parking consumes 14 percent of all space, and the city has 18.6 million parking spaces for 3.5 million housing units, at 3.3 spaces per vehicle.[14] Parking might no longer be necessary at residences, along streets, and in city centers. Unnecessary parking lots, curb parking, and road space could be converted to housing, bicycle roads, pedestrian pathways, sidewalk cafés, and parks or other types of public spaces. No longer would developers of shopping centers, apartment buildings, and places of employment be required to provide parking, or at least so much of it. Parking can also be more efficient because more vehicles can be stored in a smaller space without human access needed.

The list of potential benefits goes on. In a world where all vehicles are automated and crashes are eliminated, vehicles could be made smaller and lighter because they will no longer need heavy steel frames and safety equipment. They could also operate more efficiently because they could be programmed to drive smoothly in coordination with other vehicles, with fewer stops and starts. AVs would use a fraction of the energy required by today's vehicles and could serve as moving conference rooms. Urban environments could be decluttered through reduced signalization and signage, and fewer fueling stations would be needed.

The most pivotal consideration is likely to be personal economics. If travelers save money, market pull will be irresistible. Sharp cost reductions are not only plausible but likely—if the vehicles are operated commercially. By using the cars more intensively, rather than letting them sit unused 95 percent of the time as we now do, and by eliminating the driver, mobility providers could drive travel costs down to a fraction of what they are now—much less costly than public transit, UberPool, and all other current means of travel, as we saw in figure 1.2.

Freight also has an automated future and, in fact, might get there first. Automation will make freight transportation safer, more efficient, and less expensive. Trucks will be safer for the same reasons as cars. They can be driven in platoons, drafting each other closely and thus reducing fuel costs by an estimated 10 to 20 percent.[15] Deliveries and pickups can be optimized so that trucks are full more of the time and follow shorter routes. With partial automation, drivers will be rested and, if regulations are changed, can drive longer. Taking the driver out of the equation for automated long-haul trips lowers costs even further. The end result is safer roads and lower cost, less energy use, less greenhouse gas and pollution, and fewer trucks on the road. Not surprisingly, a mix of start-ups and longtime manufacturers are getting into the game of automated freight. These include truck start-up Otto, acquired in 2016 by Uber,[16] and Daimler, which in 2014 unveiled a prototype truck with a driverless option that it hopes to bring to market by 2025.[17] Whether laws, regulations, and the public will accept driverless trucks in a timely manner remains to be seen. Will car drivers be comfortable with walls of driverless trucks? Probably not, and thus trucks might not be driverless any sooner than cars, even though the economics are very compelling.

What Exactly *Is* an Automated Vehicle?
Automation technology can best be understood through the levels defined by the Society of Automotive Engineers (SAE). In January 2014,

the SAE established a common terminology for automated driving[18] that has been broadly embraced by government and industry. It specifies six levels of driving automation, from no automation (level 0) to full automation (level 5), based on functional aspects of the technology.

Levels of Automation

Levels 1 to 3 involve increasing use of automation technology but require a driver to be attentive.[19] As of 2017, many cars—including models from BMW, Mercedes-Benz, Audi, GM, and Volvo—use a combination of adaptive cruise control, lane keeping, and automatic braking technologies to enable drivers to take their hands off the wheel and their eyes off the road, but only briefly. This represents level-2 automation.

Level-3 cars will be able to drive themselves in some situations, such as on limited-access highways, but will need to fall back on human intervention at least occasionally. Tesla cars with Autopilot edge toward level 3. They can steer and maintain proper speed without help from the human driver. With the flick of a turn signal, they can also change lanes safely without the driver intervening. GM's Super Cruise system, made available in late 2017, allows drivers to take their hands off the steering wheel for extended periods but will stop the vehicle if drivers are not attentive.

Some companies, including Google and Ford, advocate skipping level 3, believing that these cars would be unsafe. They worry that drivers are not cognitively prepared to safely resume control of a car once it has started driving itself. Ford reported in February 2017 that the company was resorting to bells, buzzers, warning lights, vibrating seats, and shaking steering wheels to keep test engineers in the cars alert. They even put a second engineer in the vehicle to keep tabs on the main driver. Raj Nair, Ford's product development chief at the time, said that the smooth ride was just too lulling, and engineers struggled to maintain "situational awareness" (though Ford later denied they fell asleep).[20] As a result, Ford has suggested that it might skip not only level 3 but also level 4 (self-driving on limited routes with a human driver in the car), stating that it will be removing the steering wheel, brake, and gas pedals from cars it plans to make available in 2021 for use in ridesharing fleets, making these cars fully driverless (level 5).[21]

Researchers and some automakers aren't so ready to write off level 3, though. Elon Musk of Tesla insists that level 3 is an important step toward full autonomy and will save lives.[22] Shane McLaughlin, director of the Center for Automated Vehicle Systems at Virginia Tech, thinks additional technology might yet solve the human handoff problem and says, "We can get the machine to give the person enough time to react."[23] Video and infrared systems in the car might monitor the driver's attentiveness, and so-called electronic horizon technologies might give drivers more time to react by "seeing" farther down the road. But Ford has found that drivers do not appreciate sensors monitoring their facial expression and tracking eye movement to determine if they are alert and ready to take over. In test drives, drivers felt they were being constantly reminded to pay attention, complaining, "The car is actually yelling at you all the time."[24] Whether level-3 technology should be banned remains an open question.

In level-4 cars (again, fully self-driving but only on limited routes and with a driver in the car), drivers would be able to completely disengage— free to sleep, text, eat, and watch movies. Perhaps the first advanced demonstration of level-4 cars began in 2017 with up to a hundred Volvo cars operating on a limited-access ring road in Gothenburg, Sweden. While free to disengage while on that highway, the drivers needed to drive the vehicle to the ring road and knew that they might be called on to take control if something exceptional happened on the ring road— such as a diversion due to a crash.

Automated buses are a particularly attractive first application of level-4 and level-5 automation because they operate on specific, fixed, well-mapped routes. As of 2017, many companies were offering prototype automated microbuses in Europe and the United States. At present, most industry representatives and reporters are referring to level 4 when they refer to full automation or self-driving vehicles.

Still another early application could be rideshare cars operated in specified corridors or urban districts by a company such as Lyft or Uber. The corridor or district could be carefully mapped and the algorithms could assign cars only to trips beginning and ending in those carefully circumscribed areas. The cars could operate with a backup human driver for some extended period to get riders and regulators comfortable with the level-4 (and eventually level-5) cars. These early AVs would be very expensive, but the cost savings to the mobility company of removing a driver would more than offset the extra cost for the vehicle.

Level-5 cars are the Holy Grail. They will not need a steering wheel, and the passenger will not need a license—indeed, no occupant is necessary. Such cars are appropriately labeled driverless. They are much further from commercialization, in part because a large area would need to be carefully mapped and the vehicles would need to learn a lot to anticipate all surprises and also because safety regulators are likely to be conservative. The vehicles will depend on high-definition maps that let them anticipate turns and signs and recognize unmarked lanes.[25] Those maps must be much higher resolution than current GPS maps. Companies developing driverless technology will use the vehicles themselves to continually update and improve the maps, including changes in curbs and lane markings.

Vehicle Hardware and Software

Without human eyes on the road, hands on the steering wheel, and feet on the pedals, exactly how does a car plot its trajectory and avoid crashes? The answer is an integrated combination of hardware, software, and data on board the vehicle, as well as digital connections to other vehicles and roadside devices and sensors.

The key vehicle hardware consists of cameras, radar, lidar, sensors, and a controller with enormous computing power. Cameras capture images of the road, radar sends out radio waves that bounce off objects, and lidar sends out light pulses that are reflected off objects; these data are sent to the onboard computer and parsed by an algorithm looking for statistical patterns. The computer instantaneously builds a model of possible outcomes based on those patterns and instructs the car to proceed accordingly. This model gets better and better as artificial intelligence gains experience and learns from mistakes.

For a high level of safety, the car will also need to receive continuous data from traffic-control devices (and traffic police)—though that will not be available for a long time in most areas. Its onboard computer will combine geographic data with stored maps and GPS data. The stored maps need to be extremely detailed and continuously updated. But these maps and road data are not enough; traffic circumstances and the road environment are always changing. This is where the car's sensing devices—radar, cameras, sensors, and lidar—come into play. The car needs to avoid, for instance, pedestrians and dogs that are not fixed parts of the landscape. The car needs to be able to navigate through all situations and circumstances—including accepting input from humans who might direct traffic with hand motions if traffic signals fail for some reason.

Automated cars will be connected via wireless communication to other cars, roadside infrastructure, and the Cloud. Collectively, these connections are referred to as V2X (vehicles to everything). Vehicles equipped with the same operating system or with shared protocols will communicate with one another about road conditions, traffic, and landscape features. If connected with other vehicles, they will also benefit from "fleet learning," as explained by the founder of Google's automated car program, Sebastian Thrun: "When a human driver makes a mistake, he or she learns but nobody else learns. When a [connected] self-driving car makes a mistake, all the other cars learn from it as well as the unborn cars, future cars. The rate of progress is much faster."[26]

How Fast Is Vehicle Automation Coming?

Although the recent rise of automated cars might seem meteoric, the technology has been gestating for a long time. Features of automation have been integrated into airplanes and rail-transit services for many decades. Various companies and universities have experimented with automated technology in test vehicles since the 1960s. Still, vehicle automation has a long way to go before it reaches maturity, and the timeline is uncertain. The million-dollar question is, How soon can we expect driverless vehicles to take over?

The Backstory

Early forays into automated-vehicle development in the United States grew out of university engineering departments and robotics labs. A team headed by an electrical engineering professor at Ohio State University built the first AV in 1962. The onboard electronics that controlled steering, braking, and speed filled the trunk, the back seat, and most of the front passenger seat.[27] The University of California, Berkeley, developed an AV in the late 1980s that was guided by data from magnets in the roadway. In 1997, UC Berkeley hosted a demonstration in Southern California, with a platoon of eight AVs following one foot behind each other as drivers "waved their hands to show that they were not doing the steering."[28] Around the same time, in 1995, a team at Carnegie Mellon University built an automated car that completed a drive (with two researchers in the car) from Pittsburgh to San Diego in a project dubbed No Hands Across America. The project website reports that 98.2 percent of the miles were hands-free.[29]

In Europe, a consortium of European car manufacturers, electronics producers and suppliers, institutes, and universities launched a massive

R&D project in 1986 called Eureka PROMETHEUS (Program for European Traffic with Highest Efficiency and Unprecedented Safety). A number of driverless prototypes resulted, including a reengineered Mercedes S-Class car that drove almost entirely by itself from Munich to Copenhagen (1,043 miles) in 1995.[30]

The Defense Advanced Research Projects Agency (DARPA), the research and development branch of the US Defense Department, orchestrated the next big step forward. In 2004, DARPA issued the first DARPA Grand Challenge, requiring vehicles to leave Barstow, California, and autonomously navigate a 142-mile course across the desert to Primm, Nevada. None of the entrants successfully completed the course. The most successful competitor, one of two Carnegie Mellon University teams, made it only 7.5 miles.[31] The second DARPA Grand Challenge took place eighteen months later, in 2005, on a 132-mile course in southern Nevada. Five cars completed the course, and Stanford University emerged as the winner.

The third challenge, the DARPA Urban Challenge, took place in 2007 in Victorville, California. This course was different. The vehicles had to navigate a simulated urban environment, including weaving in and out of traffic and correctly passing through intersections with four-way stop signs. Six vehicles completed the course, with a team from Carnegie Mellon University taking the grand prize.

The goal of the challenges was to stimulate development of automation technologies. That they did. But the vehicles were single, hand-built prototypes, not nearly ready for commercialization. It was soon after the DARPA Urban Challenge that announcements from companies like Google and others began to pop up.

Breakthrough Firsts

If there was a turning point for vehicle automation, it was 2016. This is when the business and popular press began overflowing with adulation and enthusiasm for AVs. Major acquisitions, investments, and announcements rolled out almost daily. In April 2016, six European truck manufacturers (DAF, Iveco, MAN, Mercedes-Benz, Scania, and Volvo) entered fifteen self-driving trucks in the European Truck Platooning Challenge. Small wirelessly connected platoons of trucks (each with a human driver onboard) drove to Rotterdam from Belgium, Denmark, Germany, and Sweden—the world's first cross-border truck platooning demonstration.[32]

NuTonomy, an MIT spin-off, launched the first public AV trial in Singapore in August 2016. In September 2016, the US Department of Transportation issued the first federal policy on AVs, called by Secretary Anthony Foxx "the most comprehensive national automated vehicle policy that the world has ever seen."[33] Though a milestone, it deferred to states to promulgate specific rules. And in the same month, Uber launched a demonstration of self-driving Volvo and Ford cars in Pittsburgh. A few months later, it did the same in San Francisco (see figure 4.3), though one of us who rode in a "self-driving" Uber vehicle witnessed its unreliability navigating complex intersections and detecting bicycles alongside (with the "driver" cupping his hands around the steering wheel, poised to grab it in a split second, which frequently proved necessary).

Zoox, a Silicon Valley start-up, became the first nonautomaker or major tech company to obtain a permit to test vehicles on California roads. Drive.ai, another Silicon Valley start-up, began developing artificial intelligence software for autonomous vehicles. Many imagined Silicon Valley taking over the auto industry, with a wave of small and large technology companies selling high-profit software, dooming incumbent automakers to being suppliers of low-margin vehicle technology. Detroit and Stuttgart feared they would become subservient to Silicon Valley.

Not So Fast: Upstarts and Government Oversight
Despite their unfamiliarity with the car business, many new entrants are challenging the automotive industry. Where traditional automakers take

Figure 4.3. An Uber self-driving car prototype in San Francisco in November 2016. Photo by Dllu, CC BY-SA 4.0 (http://creativecommons.org/licenses/by-sa/4.0), via Wikimedia Commons.

five years to develop a car and operate behind closed doors on private testing tracks, the upstarts are choosing other paths. The Silicon Valley culture is to forge ahead and largely ignore or skirt government oversight and regulation. But they're finding it's not as easy as they imagined, and the reality of government scrutiny is slowing them down.

Uber is illustrative. Its Silicon Valley start-up mentality is all about being agile, moving fast, and asking for forgiveness later. Its modus operandi is to continuously launch new services on a foundation of what already exists and to outrace regulators by building strong customer loyalty and constituencies.[34] If you need expertise, buy it, as Uber did when it poached about fifty researchers from Carnegie Mellon University's renowned National Robotics Engineering Center[35] and again in early 2016, when it acquired Otto (see figure 4.4), which was cofounded by a former Google engineer. Uber stationed its automated car research centers in Pittsburgh, where Carnegie Mellon is located, and San Francisco, where Uber is headquartered. When California's Department of Motor Vehicles in late 2016 required Uber to obtain a permit to test its cars on city streets, Uber at first resisted and moved its test program from San Francisco to Arizona. But a few months later, it relented and applied for the permit.

Underlying this aggressive behavior is a libertarian belief that these innovations enhance economic growth and that individuals should be free to supply services as they please, with minimal government interference. As tech companies venture into the mobility arena, their culture is being tested. Unaccustomed to dealing with stringent regulations

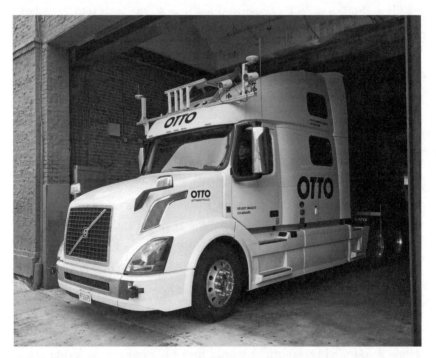

Figure 4.4. An Otto self-driving truck at the Uber research facility in San Francisco. Photo by Steve Jurvetson, CC BY 2.0 (http://creativecommons.org/licenses/by/2.0), via Wikimedia Commons.

regarding safety, they are learning that bringing a car to market is very different from launching tech gadgets and software. Unlike typical Silicon Valley technologies, which pose minimal safety risks, automated cars will receive intense scrutiny from those tasked with protecting consumer safety. Ever since Ralph Nader's book *Unsafe at Any Speed* exposed the safety risks of GM's Corvair, government oversight of car safety has been steadily ratcheted up. Needing to ensure the technology is safe, regulators impose a rigid scheme of testing, regulations, and certifications. The result is that commercialization happens at a much slower and more deliberate pace than Silicon Valley technology start-ups are used to.

In October 2015, when Tesla activated the Autopilot technology it had been building into its cars for more than a year, it simply provided a software update over the cellular network to approximately seventy thousand connected cars.[36] It treated the software as a beta version, in need of further refinement and in-field testing, though customers thought they could fully rely on Autopilot to drive the car. People posted YouTube videos of drivers asleep in the back seat. The fatal crash of a Tesla in May 2016 when it failed to detect a tractor-trailer crossing in front of it cast a harsh light on Tesla's Autopilot technology and the company's casual attitude toward safety. Tesla began instructing drivers to keep their hands on the steering wheel. The company survived the scrutiny but remained vulnerable to subsequent publicity about incidents.

Comma.ai, which aimed to give away software to make cars self-driving, pulled back after receiving a letter from the US National Highway Traffic Safety Administration. The letter explained that the company needed to ensure that its product Comma One was compliant with regulations before it could be offered for sale. The founder, George Hotz, responded by saying he would turn his attention to "other products and markets"—that the prospect of a life of "dealing with regulators and lawyers . . . isn't worth it."[37]

Recognizing the need to manage the tension between rapid commercialization and government oversight, Google, Ford, Volvo, Uber, and Lyft announced in 2016 that they were forming a coalition to "work with lawmakers, regulators, and the public to realize the safety and societal benefits of self-driving vehicles."[38]

The Uncertain Timeline Ahead

The arrival of self-driving and driverless (level-4 and level-5) vehicles is not as imminent as the hype suggests. The challenges are formidable. First, the technology is immature: cameras, lidar, and radar need to get more accurate and much less expensive; the software needs to be able to anticipate and react instantaneously and with total reliability to every scenario a vehicle can encounter; and various pieces of enabling technology still need to be developed. As Hod Lipson and Melba Kurman observe in *Driverless: Intelligent Cars and the Road Ahead*, "The technological last mile in driverless-car design is the development of software that can provide reliable artificial perception. If history is any guide, this last mile could prove to be a long haul."[39] A lack of visible lane markings and reliable traffic signals in many areas only makes the challenge more difficult. For reliably safe vehicles, vehicle-to-vehicle and vehicle-to-infrastructure communication systems need to be developed, though they might not be essential for safe operation during a transition period.

Assuming the technical challenges can be met, how soon will driverless vehicles serve the majority of travelers? Probably not for a long time—several decades at least, even in ambitious scenarios. Consider that people have huge sunk costs sitting in their driveways, with current vehicles lasting as long as twenty years and sometimes more. There are nearly three hundred million vehicles in the United States, with a turnover of around seventeen million annually. If every single vehicle sold from 2018 forward were driverless, it would take until 2034 for all the cars on the road to be driverless.

One analysis, based on the commercialization history of previous vehicle technologies such as airbags and automatic transmissions, suggests that it will take one to three decades for AVs to dominate vehicle sales and one or two more decades to dominate vehicle travel. This analysis suggests that if technical challenges prove to be more difficult to solve than expected, fully self-driving vehicles might not be commercially available until the 2030s or 2040s.[40] And if public and regulatory acceptance plays a role, it will be even longer.

Our more conservative views are reinforced by surveys of consumers. For AVs to be widely adopted, consumers at least need to be willing to ride in them. A 2015 survey of licensed drivers in the United States conducted by University of Michigan researchers found that consumers are intrigued by automation, but not full automation. The most preferred vehicle automation was, ironically, no driverless capability. Only 15 percent of participants preferred completely driverless vehicles to partially or non-self-driving vehicles. And nearly all participants answered yes when asked if they preferred that the vehicle "have a steering wheel plus gas and brake pedals."[41] A survey by the Insurance Information Institute in May 2016 found that 55 percent of respondents would not ride in an AV.[42] Another study about the same time by the Boston Consulting Group, in collaboration with the World Economic Forum, found somewhat more open attitudes, with most people saying that an AV was appealing because they would not need to find parking and would be able to multitask while riding, but again they rated their top two concerns as not feeling safe and wanting to be in control.[43] Of course, people tend to be skeptical and cautious about something they've never experienced. That skepticism could evaporate quickly as AVs become commonplace. People were afraid of automated elevators when they were first installed.

Car companies have mixed views about whether humans really want to hand over full control of their cars, and even if they are willing, companies waver in their aspiration to that goal. Until now, cars have been designed to be driver centric: the focus of automotive design is the front seats, and especially the driver's seat. The current business approach is to add automation features to cars incrementally and to treat those increments as premium add-ons. This approach is especially appealing to makers of luxury cars, who can charge high premiums for technology enhancements. Indeed, both Mercedes and BMW have kept the steering wheel in their extravagant future concept cars. If they accede to driverless cars, with the same basic (safe) technology embedded in all cars, how do they distinguish themselves from others? For many reasons, it looks to be a long leap to a passenger-centric car that excludes a steering wheel and brakes.

What Are the Potential Downsides of Vehicle Automation?

The full embrace of driverless cars could transform cities, lifestyles, and vast swathes of the economy. The reverberations could be so broad and deep that vehicle automation could rank in history as a true industrial revolution, following on the heels of the information revolution.

Singapore: Bullish on Autonomous Technology

Singapore is one of the most affluent countries in the world and a major air and shipping hub. Its 5.5 million people live on a small island of 278 square miles,[44] with roads covering 12 percent of the land area[45] (and more for parking). Singapore has an urgent need to find mobility solutions that do not require more land, serve its affluent society, and are economically feasible and environmentally friendly. An MIT study of Singapore in 2014 suggested that an automated mobility system in the city could reduce passenger vehicles to a third of current numbers.[46] The government has concluded that a combination of new technology, new business models, and forward-thinking regulation is necessary to meet its needs.

Singapore is a world leader in limiting vehicle travel, going back to a road-pricing scheme introduced in 1975 that required a fee payment to enter the most congested parts of the central business district during morning peak hours.[47] Singapore now has in place a vehicle quota system, hefty vehicle

---

purchase and registration fees, traffic congestion charges, an extensive public transport system that offers free rides in the morning peak hours, and intelligent traffic signals that are networked and responsive to traffic.

In 2014, the Committee for Autonomous Road Transport for Singapore (CARTS) was formed to "provide thought leadership and guidance on the research, development, and deployment of AV technology and AV-enabled mobility concepts for Singapore; and study the associated opportunities and challenges."[48] CARTS—which includes international experts, academics, and industry representatives—imagines a future in which people can get wherever they want, whenever they want, without the stress of driving, on roads that are safe and in neighborhoods with green open spaces, and where the elderly and disabled have enhanced mobility. What they call "AV-enabled mobility" could help commuters by offering first- and last-mile service and reducing congestion on the roads.

Also in 2014, Singapore's Land Transport Authority (LTA) signed a five-year collaboration with the city's lead research and development agency to launch the Singapore Autonomous Vehicle Initiative to explore technological possibilities for AVs. Some R&D trials were already under way: MIT was partnering with the National University of Singapore to test a fleet of driverless golf buggies on campus, and a driverless shuttle bus was running from the Nanyang Technological University campus to Cleantech Park. Starting in 2015, the city issued permits to four agencies to use the One-North business district, a 1.5-square-mile section of Singapore, for testing driverless vehicles. The LTA also signed partnership agreements with MIT spinoff nuTonomy and Britain-based Delphi to test self-driving taxis in Singapore.

In September 2016, nuTonomy joined forces with Grab, an Uber-like company, to offer free driverless, on-demand trips in a public trial in One-North. Members of the public can use the Grab app to hail a ride in an automated electric vehicle manufactured by Renault or Mitsubishi.[49] The company had its first accident in October 2016, when an automated car ran into a truck and caused minor damage, said to be due to "an extremely rare combination of software anomalies." But after software fixes, nuTonomy cars were back on the roads, with the company saying the incident demonstrated "the value of conducting extensive testing," something Singapore's policies have made possible. The company hopes to provide limited commercial service in 2018 and roll out the services islandwide by 2020.[50]

Combined with shared mobility and vehicle electrification, the transformation will be even deeper and reach even further.

But the benefits many envision are not guaranteed. Until most vehicles are fully driverless, the benefits will be modest, including safety. Mixing driverless and nondriverless cars will have uncertain safety benefits. And many of the potential benefits for the mobility disadvantaged will also not be realized until fully driverless level-5 vehicles are available. When fully driverless cars are widely available, benefits will result, but they will accrue only to a minority of rich people unless the rides are shared. Only if driverless vehicles are shared—as opposed to individually owned—will the large benefits of reduced vehicle use, greenhouse gas emissions, and traffic congestion be realized. Other concerns about self-driving (level 4) and driverless (level 5) cars include issues with trusting robots to make choices about life and death, hacking of cars, privacy loss, and job losses.

Safety Concerns and Traffic Congestion

Partially automated cars—those with adaptive cruise control, emergency braking, and some steering assistance—have some safety benefits, mostly coming from the emergency braking, which is already widely used. The first safety concerns come with level-3 automation, whereby cars could be self-driving on freeways and perhaps elsewhere but would require a return to human control when unusual situations arise and when traveling on other roads. These might turn out to be less safe and cause more crashes.

Moving upward to level-4 self-driving and level-5 driverless cars will eventually generate huge safety benefits, but those benefits might not be realized for a long time because of the challenges of mixing driver-operated and driverless vehicles on the same roads. Mastering responses to the unpredictable actions of human drivers is challenging. Hardware and software will need to figure out how quickly cars are merging into freeway traffic, anticipate when someone is about to pull a U-turn or make an illegal left turn, and myriad other actions that can result in deadly crashes. Under almost any scenario, it will take many decades for the full transition to driverless cars to be realized. In the meantime, much care will need to be taken to ensure that mixing human- and computer-driven cars results in increased safety for all.[51] Eventually, though, self-driving cars will significantly reduce traffic fatalities, and for that reason, vehicle automation is highly compelling.

Whether AVs improve or worsen traffic congestion depends on the levels of automation and adoption.[52] Partially automated cars are unlikely to improve traffic congestion substantially and could even worsen it by easing the driving burden and encouraging drivers to seek cheaper homes in the exurbs and thus drive more.[53] There might also be a stretching out of distances between vehicles. Those of us with adaptive cruise control know that our partially automated cars tend to leave larger gaps than we do as human drivers (at least in large cities like Los Angeles and New York, where drivers are experienced with congested and high-speed driving). Even level-4 self-driving and level-5 driverless cars are not likely to improve traffic congestion as long as they are mixed with regular vehicles. "Self-driving cars won't break the law and will be inherently cautious, which means the presence of even one in a mixed autonomous/human driving environment will reduce speeds," asserts race driver Alex Roy.[54] Given this reality, and the ease of driving, it's possible that during the decades when driverless cars are not the majority, urban congestion could increase.

Driverless level-5 vehicles could also worsen traffic during the transition period if they are personally owned. If owners of fully automated vehicles send their empty cars back home (where parking is free) after the morning commute and then request their empty cars to pick them up at the end of the day, this will essentially double the vehicle miles traveled for that commute. While this type of behavior will depend on energy prices and parking costs, for some commuters this could be the lowest-cost solution to parking—or at least it might seem to be, based on a simple analysis of out-of-pocket costs. All else being equal, making vehicle travel more convenient and more possible for nondrivers and seniors will result in increased vehicle use.[55]

Issues with Putting Robots in Control

Who is responsible for an AV crash? It's not the driver! Drivers might not be terribly concerned with the distinctions between different levels of automation, but insurance companies most assuredly will be. Car insurance is expected to change radically in the era of self-driving cars.

The US National Highway Traffic Safety Administration (NHTSA) has determined that the driver of an automated car is its computer. If this is contested in court, the original manufacturer of the vehicle and its equipment suppliers will inevitably lock horns over who is responsible for crashes and equipment faults. In many states, approval of permits to test vehicles requires operators to demonstrate that they are capable of covering any damages caused by the AV up to a certain dollar amount. The party responsible for this coverage is the developer and tester of the technology. In some cases, this might be the auto manufacturer, but in cases where the self-driving capabilities are added to a vehicle after its original sale, it might be the after-market supplier.

Stepping back further, what decisions should the automated car be making? Imagine that an automated car detects pedestrians it can't avoid and needs to make an instantaneous choice between running them down or swerving and crashing into a tree, possibly killing the occupants. How does one program an AV to make ethical choices like these? William Ford Jr., executive chairman of Ford Motor Company, has asserted that automakers and regulators are going to have to work together to decide such things. "Could you imagine if we had one algorithm and Toyota had another and General Motors had another?" Ford asked at a meeting of the Economic Club of Washington, DC, in October 2016. "We need to have a national conversation about ethics, because we have never had to think about these things before, but the cars will have the time and ability to do it."[56]

The NHTSA has asked automakers to submit safety assessments that explain how their automated cars would handle ethical dilemmas. Christoph von Hugo, manager of driver assistance systems at Mercedes-Benz, told *Car and Driver*, "If you know you can save at least one person . . . save the one in the car."[57] At the same time, he made clear that "99 percent of our engineering work is to prevent these situations [that pose the moral question of whom to save] from happening at all." He later changed his mind about prioritizing passengers, highlighting how uncertain automakers are about these moral decisions.

The right way to program automated cars is not at all clear. One survey of consumers found that most thought the moral choice for the car would be to sacrifice itself for the greater good, but at the same time, most of the respondents said they would prefer not to ride in a car programmed this way. The study authors concluded that regulations designed to force manufacturers to program AVs to sacrifice their own passengers "may paradoxically increase casualties by postponing adoption of a safer technology."[58]

One legal researcher suggests this ethics debate might be moot. He asserts that lawyers and the market will essentially tell engineers how to program the cars.[59] Lawyers will determine liability rules. Vehicle suppliers will be "less concerned with esoteric questions of right and wrong than with concrete questions of predictive legal liability." And consumers will make clear what they want, as when they made clear to Tesla that they wanted their Autopilot cars to drive above the speed limit. Tesla accommodated. Software engineers, according to this view, will accommodate lawyers and consumers.

Cyber Security Risks and Privacy Concerns
Even more troubling is the risk of hacking. As vehicles become more like computers and robots, they become more susceptible to hackers taking control. In 2015, two hackers with a laptop remotely sent commands to

a Jeep Cherokee that turned on the air conditioning, switched the radio station, turned on the windshield wipers, and stopped the accelerator pedal from working. It was just an experiment, but they proved that anyone with the right code can gain wireless control of thousands or even millions of vehicles.[60] In 2016, Chinese cyber security researchers remotely tapped the brakes and popped the trunk of a Tesla Model S. The CIA has even shown interest in vehicle cyber security, citing "vehicle systems" as "potential mission areas" in a document leaked by WikiLeaks in early 2017.[61] Imagine traffic signals as well as vehicles being hacked, and you have a recipe for transportation chaos. This scenario is possible with the rise of smart machines, high-speed wireless connectivity, big data, artificial intelligence, and cheap sensors.[62]

The bottom line is that the software driving AVs will be ripe for hacking and sabotage. Indeed, millions of vehicles on the road today are already vulnerable because they are connected to the Internet. As noted by one cyber security firm, "Once a vehicle connects to the Internet, it is hackable. . . . A vehicle has multiple penetration vectors, with 100 million lines of software code and an average of 10,000 known software bugs in it when it rolls out of Detroit or Stuttgart."[63]

This is a software and hardware engineering problem, and we can hope it has an engineering solution. QNX, which makes the operating system used by most global automakers, has pointed out that given the collection of hardware, software, and network components that make up a connected car, "security is only as strong as its weakest link."[64] A key strategy is to reduce the number of gateways for communication with crucial systems and to require services offered by third parties to use a single secure path, but the industry is years away from solving the cyber security problem.

An associated issue is privacy. With their high-definition cameras and sensors, automated cars could track people both inside and outside the car and send information to intelligence agencies, law enforcement, and others. Cyber criminals might try to penetrate a vehicle system to steal personal information or determine a driver's location. Again, this is a solvable problem, but much effort will be needed to align engineering solutions and stakeholder interests with legal and liability solutions.

The challenge is to create rules and laws that are effective at blocking hacking and protecting privacy but are not prescriptive. The goal should be to stimulate innovation, not quash it.

Loss of Jobs—or Creation of More Jobs?

Automation means fewer jobs, as a first-order generalization. Machines replace workers—in this case, drivers. But that simple proposition might not be correct in this case, certainly in the near and medium term but even in the long term. In the near term, partial automation (levels 1 to 3) will have little or no effect on jobs. That will be largely true of level-4 automation as well. It is more complicated for level-5 driverless cars. If most of these cars are individually owned, there will be little effect because no workers will be displaced.

If automation is used for commercially owned driverless (level-5) cars and vans, paid drivers will be displaced, but even so, the net effect may be positive. Consider that current taxi, bus, and ridehailing services in the United States carry only about 2 percent of all passenger travel.[65] We are imagining a future where driverless mobility services and buses account for 10 or 20 percent or even a much greater share of passenger travel. With such a mammoth increase in commercial ridesharing and transit use, some jobs will be displaced but far more will be created—to service the vehicles, manage the fleets, assist disabled riders, and handle untold other tasks. In effect, AVs will be creating a new industry, not just replacing an old one. They will be creating jobs to replace previously unpaid personal driving and vehicle maintenance, not unlike what happened with mechanization and automation of cooking and other household tasks. The number of lost driving jobs might well be dwarfed by the number of new jobs created—and they might well be higher-quality and better-paid jobs.

Sebastian Thrun, a pioneer of self-driving vehicles, makes this point: "There are a million or more people whose jobs might change or be in jeopardy if self-driving cars become a phenomenon of scale, found everywhere—which might take ten or twenty years. Taxi drivers, insurance brokers, trauma surgeons, tort lawyers. But the beauty of innovation is that new jobs open up as we develop technology to get rid of repetitive and mindless human labor."[66]

For goods movement, the implications for labor are less clear. In a worst-case scenario for workers, says a report from the International Transportation Forum in Paris, "automated trucks could reduce the demand for drivers by 50 to 70 percent in the United States and Europe by 2030, with up to 4.4 million of the projected 6.4 million professional trucking jobs becoming redundant."[67] However, as the report goes on to say, "fewer than 5.6 million [truck drivers] are projected to be available and willing to work under current conditions. The majority of truckers are in the later stages of their careers, while few women and young men are choosing trucking as a profession."[68] Moreover, there is almost no chance the transition will be rapid. And as noted previously for passenger transport, truck automation will generate many other jobs, which might prove more attractive than today's trucking jobs.

How Can We Steer Automated Vehicles toward the Dream Scenario?

The transition to driverless vehicles is inevitable. The technology is inherently attractive to individuals and businesses because it is safer and saves time. Many people will pay a lot of money to acquire AVs, both cars and trucks. Automotive and technology companies have every reason to meet this demand. They will be providing a higher-value product that sells at a higher price (and, they hope, higher profit). Market pull will be powerful.

But how do we direct the transition in a way that benefits cities and society? Integrating pooling, vehicle electrification, and automation is essential. A key challenge right from the beginning will be to encourage pooling—as Lyft and Uber have done since 2014 with Lyft Line and UberPool—and then discourage individual ownership of automated cars. The market by itself might push vehicle automation toward shared and electric vehicles, or it might not. Either way, government at every level can accelerate the process, nudging consumers and automakers toward pooling and away from personal ownership of automated cars.[69]

At the international level, there would ideally be an international entity that would harmonize vehicle standards around the world. That is unlikely well into the foreseeable future—just as it is unlikely that international bodies will regulate air pollution, energy use, and greenhouse gas emissions.

At the national level, the primary focus of governments should be the safe design of vehicles, leaving lower-level governments to focus on the use of the vehicles (though inconsistent state regulations pose a risk[70]). A first step in the United States was the September 2016 publication of the Federal Automated Vehicles Policy, which provides guidelines on safe testing and deployment, a model state policy, and prospective regulatory tools,[71] but much more specific rules and guidelines on safety, privacy, and cyber security will be needed. National governments can also provide funding to guide the evolution of urban transportation services as automation is added, though most transportation funding in the United States and many other countries comes from state and local levels.

State and local governments will be at the forefront. They deal with operation of vehicles, including licensing of drivers and vehicles, and management of roads and parking. As of early 2017, nine US states had enacted AV legislation, setting rules on where the vehicles can be operated and whether they need a driver in the vehicle. More challenging is developing policies and incentives to guide which vehicles can and should be used where. Few states have done so as of 2017 because of great uncertainty about how the technology will progress, what consumer responses will be, what federal policies will be adopted, and what exactly the goals should be.[72] Robin Chase, cofounder and former CEO of Zipcar (and coauthor of chapter 1 in this book), has suggested that cities sign on to a document entitled "Shared Passenger Mobility Protocol for Livable Cities" and thus form a collective bargaining bloc to deal with the large multinationals that will be selling their vehicles and services.[73]

Currently, policy makers and public agencies are struggling to catch up with the torrent of new technologies and services. Their very difficult challenge is to protect the public interest without stifling innovation. In the best case, policy solutions can support the most socially and environmentally beneficial aspects, discourage the potential downsides, and address the potentially disruptive outcomes of AV services and technologies. In practice, it will be a messy process, and it will take time. But now is the moment to start shaping how the automation revolution will transform transportation—how vehicles and mobility services will be used, who they will benefit, and how they will impact the environment.

Key Policy Strategies

The aim is to give travelers access to low-cost and environmentally sustainable mobility by encouraging automakers to power AVs with electricity and hydrogen and motivating mobility service companies to use these vehicles for pooling.

At the National Level
- In regulations for fuel economy, greenhouse gas emissions, and zero-emission vehicles (ZEVs), award extra credit to automakers for level-5 AVs used for pooling services.
- Define a new Federal Motor Vehicle Safety Standard (FMVSS) to allow level-5 features, such as omitting driver controls, with explicit accompanying safety requirements.

- Make AV technologies eligible for federal safety grants programs to improve transportation operations.
- Require the auto industry to protect the privacy of vehicle owners while enabling mutually beneficial data sharing agreements between private and public entities.
- Set operation requirements for AVs to drive efficiently, at safe speeds, and with minimal braking.
- Update laws that prohibit and punish any tampering with or disabling of AV communications.

At the State or Local Level
- Incentivize operators of shared commercial fleets to purchase and use level-4 and level-5 AVs that are powered by electricity and hydrogen.
- In regulations for ZEVs, award extra credit to automakers for level-5 AVs designed for pooling services.
- Impose low (or zero) registration fees on level-5 AVs used for pooling and high fees on AVs purchased by individuals.
- In very dense downtowns, strongly discourage zero-occupancy and single-occupancy vehicles by requiring special permission or imposing large disincentives.
- Impose high tolls and fees on zero-occupancy and single-occupancy vehicles operated on freeways (and eventually on all roads).

CHAPTER 5

Upgrading Transit for the Twenty-First Century

Steven E. Polzin and Daniel Sperling

Urban transit has struggled for decades. The oncoming disruptions could invigorate public transportation and transit agencies—or lead to their decline.

In the United States, cars have increased their dominance over public transit. Overall, transit accounts for only about 2 percent of passenger trips in the United States[1] and about 1 percent of passenger miles.[2] Even in the developing world, transit has lost market share to cars. Public transit is struggling. As the world changes, public transit needs to reinvent itself.

Support for public transit endures. Voters in many large American cities continue to approve large subsidies. In November 2016, Los Angeles voted to raise sales taxes to provide $120 billion mostly for bus and rail services over the next forty years. Since 2000, more than two hundred localities have voted to tax themselves to finance improvements, almost always featuring public transit. The ballot success rate for such measures is greater than 70 percent, far higher than the success rate for tax referenda overall.[3] Citizens clearly believe public transportation provides economic, energy, environmental, and mobility benefits. This broad political support has not, however, translated into higher ridership.

Now comes the boom in shared mobility and automation technologies. Will it be a boon or a curse for transit operators? Some predict the demise of public transit as we know it, with automated on-demand vehicles replacing traditional fixed-route services in all but the highest-volume corridors.[4] The media are replete with provocative headlines such as "Super-Cheap Driverless Cabs to Kick Mass Transit to the Curb."[5] Some articles are more thoughtful, asking "Is Uber Killing the Public Bus, or Helping It?"[6] The most hopeful stories ask how new mobility services can feed riders to beleaguered transit operators.[7]

There will be some disruption, but how much is uncertain. What *is* certain is that travelers will have more choice. They will be presented with an array of new options varying by cost, speed, convenience, flexibility, safety, reliability, comfort, and environmental impact. Can transit operators remain competitive? Doing so is in the interest of all that they do.

The Decline of Public Transit

Mass transit plays a valuable role in virtually all economically advanced countries and an even more vital role in emerging economies and less-developed countries. But with the rise of the automobile, first in the United States and then elsewhere, transit ridership has suffered. Its decline has been most precipitous in the United States—not coincidentally the land where cars most dominate. Transit supply has grown dramatically, nearly doubling since 1970 (measured as vehicle miles of service), while ridership has increased only about 42 percent, slower than population growth (see figure 5.1).[8]

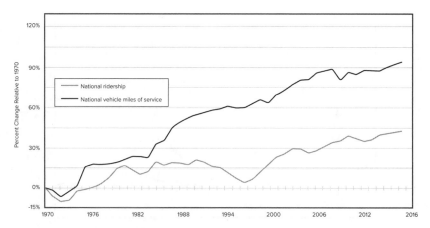

Figure 5.1. Change in US transit supply and ridership (excluding demand-responsive, transit vanpool, fixed-route taxi, and ferry in both) relative to 1970.

Ever since the advent of the Model T, US public transit patronage has been under pressure. The only exception was during World War II, when gasoline was rationed and new cars were not available.[9] After the war, as car ownership, suburban growth, and highway investment soared, transit ridership fell from 114 trips per capita in 1950 to 31 in 1973 and then stabilized as transit subsidies increased (see figure 5.2; total trips increased slightly due to population growth, but because people traveled more, transit's market share continued to drop slightly). Trips most amenable to transit, work trips in urban areas, have been falling dramatically. Transit carried 12.5 percent of work trips in 1960, dropping to 8.5 percent in 1970, 6.2 percent in 1980, and then only about 5 percent in 1990 and beyond.[10]

Even as ridership stagnates, subsidies have continued to increase. Controlling for inflation, total local, state, and federal subsidies to transit doubled between 1988 and 2010.[11] Today, urban bus passengers in the United States pay only about 20 percent of the full cost of the service (rail riders pay 30 percent).[12] The other 80 percent (70 for rail) is covered by government subsidies.

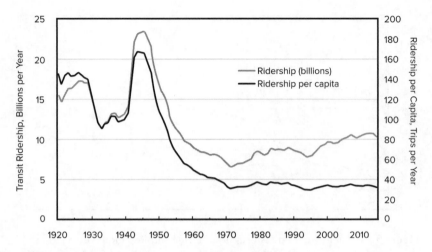

Figure 5.2. US transit ridership trends, 1920–2014. Source: Compiled from the fact books published annually since 1942 by the American Public Transportation Association and from US Census data.

The only positive note for transit is significant new investments in rail. Between 2000 and 2012, nearly twenty cities in the United States introduced or expanded rail service, resulting in light-rail passenger miles rising 90 percent and heavy-rail miles rising 14 percent.[13] Still, transit agencies are struggling to cover costs even as subsidies increase. The result is deferred maintenance, fare increases, service cuts, and underfunded pensions.

It is difficult to imagine a scenario in which public transit as it exists today would significantly expand ridership. The cost of expanding service to attract more riders is huge, especially in suburban areas. The average cost of providing bus service in the United States is nearly $5 per trip.[14] The cost of attracting new trips in marginal markets would be much higher.

Looking to the future, the transit industry is challenged with finding ways to provide service for riders and regions that are unable to afford traditional public transportation services. Partnering with private companies that offer technology-enabled services is a tantalizing way to expand travel choices without incurring the high cost and inconvenience of traditional fixed-route services.

The Political Challenge of Change

Transit has a broad constituency even though it serves relatively few people. While the case for new transit partnerships is compelling, the transition is fraught with political land mines. Transit agencies employ union labor, offer low fares so that they can serve low-income riders, and often extend routes into low-density (wealthy) suburban communities, often at very high cost. While these actions increase costs, they also boost public support. Shutting down these lightly used routes, hiring nonunion companies, and raising fares would likely undermine support and invite political backlash.

No matter how appealing and rational partnerships with private operators might be, the benefits will be uneven and specific to location, threatening the relatively broad support of public transit. If a new worldview emerges that public transit operators are not principally the servant of disadvantaged travelers, the social compact could unravel. Support for public transit could be challenged almost everywhere except large cities, and even there, it might be threatened. Those large subsidies could wither away.

Lyft and Uber are already undermining transit ridership. Preliminary data suggest that although the two ridehailing companies are delivering some travelers to transit stations, they are also diverting trips from public transportation. The diversion in New York and San Francisco appears to be substantial—with 10 to 30 percent of Uber and Lyft riders switching from transit.[15] Automation would exacerbate the risks to transit operators because shared automated cars and vans could greatly reduce costs, possibly as low as (subsidized) transit fares. Attractively priced automated car and van services would be an especially large threat in markets where transit service is not time competitive, transit routes are circuitous, or transit is not convenient or comfortable (for example, inadequate seats and steep steps).

These new services, automated or not, will siphon off higher-income travelers who can afford them. The result is fewer transit riders, per-petuating the downward spiral in which lower productivity and lower fare revenues lead to additional service cuts that in turn lead to even lower ridership. If the constituency for public transportation shrinks—particularly if it loses influential suburban customers—this could under-mine political support for public transportation spending. Low-cost pooling and automation services could also lessen the appeal of transit-oriented developments[16] by reducing price premiums for housing located near transit stops.[17]

In richer countries, transit is called on to address a long list of often intractable problems. The American Public Transportation Association, the trade association for US transit operators, asserts that transit creates jobs, raises tax revenues through the multiplier effect, saves individuals money on travel, gets cars off the road, reduces reliance on foreign oil, increases business sales, reduces greenhouse gas emissions, reduces sprawl, saves large amounts of travel time, aids the disabled, and improves public health by promoting walking.[18]

This something-for-everyone political strategy is problematic. It results in not accomplishing any of the goals well and leads to high costs. To accommodate low-income riders, fares are kept low. To accommodate disabled riders, expensive "kneeling" buses and paratransit services are provided, and space is set aside on board for wheelchairs. To retain polit-ical support, service is extended into wealthy suburbs for commuters and housekeepers, often at very high cost. If transit agencies were businesses, they would operate very differently. But they aren't businesses because we as a society have committed to providing a minimal level of service to the most vulnerable among us. In short, transit's goals, though individually worthy, in concert translate into high costs and thus sparse service.

The safety-net argument for subsidizing mass transit remains valid and strong. The same cannot be said for the environmental argument, especially for buses, at least in the United States. Because transit supply has increased over time while ridership has not, the average occupancy of buses and trains is low. With cars getting more energy efficient, buses do not fare well in comparison. In 1990, cars for the first time became more energy efficient (and emitted less greenhouse gases) than buses on a passenger-mile basis. Cars are now more than 20 percent more efficient than buses.[19] Cars and light trucks—minivans, SUVs, and pickup trucks—together emit less greenhouse gases and use less energy than buses. The gap is increasing as light-duty vehicles become more energy efficient.

The greenhouse gas intensity of rail transit is better, about half that of buses when only counting propulsion energy,[20] and is aided significantly by the high productivity of the mature systems in large, dense cities. But if we analyzed the full life-cycle emissions, including construction of tracks and stations and energy expended in rail and station upkeep, the gap would be much smaller. If greenhouse gas reduction is the goal, the cheapest and easiest strategy is to reduce the carbon footprint of cars, not increase the occupancy of buses.

It is in this risky and uncertain terrain that transit is challenged to carry out its core mission of providing affordable mobility for disadvantaged persons and high-capacity mobility in dense corridors.

Although the political challenge for transit operators is large, moving forward is less risky than standing still. Until now, criticism of transit has been muted, even as subsidies have mounted and ridership has stagnated. The high energy and financial cost of public transit has received little attention. Criticisms have been deflected by focusing on potential benefits, such as densification around stations, economic development, and urban livability. Criticism has been muted until now because there were no good alternatives. Now that mobility alternatives are beginning to emerge, criticism will no longer fall on deaf ears.

Transit operators need to engage the mobility revolution if they are to survive, much less thrive. Reenvisioning transit for the twenty-first century will require integrating and leveraging new technologies and service models. Transit agencies can resist, but the cost of doing so could be marginalization and even extinction.

New Service Models and Partnerships to Bolster Transit

In 1994, Professor Melvin Webber of the University of California, Berkeley, published a prescient article, "The Marriage of Transit and Autos: How to Make Transit Popular Again."[21] He proposed using computer-enabled rematching to fill small passenger vehicles so that transit could compete with cars in terms of cost, convenience, and wait times, even in suburbs. He suggested that "the ideal suburban transit system will take its passengers from door to door with no transfers, with little waiting— and . . . it will fit the small number of persons having the same origin, the same destination, and the same schedule." This, he believed, was the only way to provide mass transit that could compete with the private auto and reverse the long decline in transit ridership.

There are limits to his vision. Smaller vehicles—vans and cars—cannot efficiently replace high-volume services in denser cities. Conventional services with fixed routes and fixed schedules will always remain— although they would be more competitive if outfitted with automated controls that eliminated drivers. Fully automated people movers are already in use at many airports and in some city downtowns, exemplified by the Metromover in downtown Miami; the Morgantown Personal Rapid Transit system in Morgantown, West Virginia; and the Las Vegas Monorail. Such automated trains, along with express buses operating in high-density corridors, sometimes with reserved lanes (called bus rapid transit), would provide an even stronger backbone for public transportation. These fixed-route services carry large numbers of passengers and will continue to be offered by legacy transit operators.

But in suburbs, areas away from rail lines, and other communities not well served by fixed-route bus services, demand might better be addressed with smaller vehicles and more flexible routing and scheduling, as suggested by Webber. New service models and partnerships are already emerging to serve shorter trips to and from the line-haul routes. Others will emerge to serve the mobility disadvantaged—those unable or too poor to own a car. To the extent that they are less expensive than fixed-route, fixed-schedule bus services, new models provide an appealing option to reduce costs.[22] Ultimately, automated shared-ride vehicle services will reach far-flung people and places, transport persons with disabilities, plug first- and last-mile gaps, and feed into public transport trunk lines.[23]

Transit agencies around the globe are experimenting with a variety of new service concepts. The numerous demonstration and pilot projects under way and in the planning stages include partnerships with app-based demand-responsive services such as Lyft and Uber, real-time rideshare-matching services, and short-term car rental and bikeshare services. Several transit agencies in the United States and Canada are subsidizing Lyft and Uber in lieu of offering (more expensive) traditional bus service. The Los Angeles County Metropolitan Transportation Authority, Bay Area Rapid Transit (San Francisco), and Metro-North/MTA of New York City have positioned bikesharing and carsharing facilities at bus and rail stations. Kansas City's transit agency partnered with a microtransit provider to test an extension of its service.

These many pilot projects provide a growing body of experience that can be the basis for planning better public transit connections and services. Of particular interest will be discerning the market contexts in which first- and last-mile services can increase ridership for fixed-route services. Ultimately, transportation providers will need to understand the cost, performance, and environmental impacts of various combinations of services. This will include determining what level of investment

Subsidizing Ridehailing in Altamonte Springs and Pinellas Park, Florida

Large transit subsidies keep fares low. Fares paid by transit riders in the United States cover less than a third of the operating cost and none of the capital cost. Riders pay an even smaller percentage in less dense suburban areas.

Now a couple of transit agencies in Florida are experimenting with replacing inefficient and expensive bus lines with subsidies for riders who use Uber. In 2016, Altamonte Springs, a suburb of Orlando, began paying for 20 percent of any Uber ride within city limits, a rate that ramps up to 25 percent for rides that end at regional rail stations. The program has grown quickly and is expanding to several neighboring towns.

The Pinellas Suncoast Transit Authority in Pinellas County, Florida, began a similar program after voters rejected a referendum in late 2014 to raise taxes to pay for more buses and a light-rail system. The transit operator responded with plans to cut bus lines to Pinellas Park, a relatively dense working-class area, and East Lake, a more affluent area with a publicly run van service. When residents complained that they would be stranded, the transit agency started a pilot program to subsidize Uber rides for anyone who would have traveled those two routes. Covering 50 percent of riders' fares turned out to cost only a quarter of what it took to run the two bus lines.[24] In August 2016, the agency expanded the initial program to subsidize all Uber rides that end at a couple dozen designated transit stops. It also began giving free Uber rides to low-income residents traveling after 9:00 p.m., when buses aren't running.

Despite questions about how to serve people without smartphones, how well ridehailing services such as Uber and Lyft serve disabled riders, whether governments should encourage replacement of public-sector jobs with contract work, and whether ridehailing companies are actually in it for the long haul, this partnership model of service seems likely to grow. The experiments continue to spread. Innisfil, a small but sprawling Canadian town north of Toronto, inaugurated a pilot program in May 2017 to partner with Uber to provide subsidized transportation for the town's thirty-six thousand residents rather than develop a more traditional public transit system.[25] Centennial, Colorado, is partnering with Lyft to provide rides to and from a regional rail stop. Miami-Dade County wants to subsidize Uber and Lyft rides to two train stations. And Uber has struck public transit agreements with San Francisco, Atlanta, Philadelphia, Dallas, Cincinnati, Pittsburgh, and others.[26]

in complementary services is prudent relative to traditional strategies such as reduced fares, more park-and-ride lots, increased frequency, more routes, and expanded hours of operation.

Going forward, the public transportation industry—and local leaders—will have to assess the ability of ridehailing companies to be good partners and provide reliable service, adequate capacity, and stable pricing.

Partnering with Microtransit Providers

Another partnership opportunity—beyond Lyft, Uber, and other ridehailing services that use personally owned cars and SUVs—is microtransit companies that operate vans and small buses. Microtransit encompasses a broad mix of demand-responsive curb-to-curb services. Many private providers have emerged in recent years, though with mixed success.

The Kansas City Area Transportation Authority (KCATA) partnered with a private company, Bridj, to provide on-demand microbus service as a way to extend the capabilities of Kansas City's existing public transportation system.[27] Launched in March 2016, this partnership was the first of its kind, bringing together a major urban transit agency with an automotive manufacturer and a private, technology-based shared mobility company. The program utilized Ford Transit vans for the service and allowed travelers to hail the vans through the Bridj mobile app, which then crowdsourced the ride hails and created a virtual meeting point within the designated service area. Fares for the service were subsidized so that they were no more than the standard bus fare.

It failed. Only 597 rides were given in the first six months of the pilot project, a tiny fraction of the 200 riders per day projected at the program's outset.[28] Staff at the government transportation agency said poor communication to the public was the principal explanation for poor ridership. An independent survey of riders found that residents didn't find the service particularly useful.[29] Nearly half of those who requested a ride used the service only once, despite introductory promotions. Of the people who downloaded the Bridj app but never used it, three quarters indicated this was because it didn't serve locations where they wanted to go. A third also responded that it didn't operate when they needed it. The KCATA on-demand pilot program ended, and Bridj closed its doors in May 2017, as noted in chapter 3. The concept of serving more-dispersed travelers with small on-demand buses is compelling, but in this case, the execution was lacking.

LA Metro, the transit provider for the Los Angeles area, is determined to learn from those mistakes. It is proposing to offer microtransit services to complement its rail and bus offerings. The agency cites many advantages of microtransit: flexibility to reroute in real time, scalability to meet demand as it changes throughout the day and week, ability to serve a greater variety of origins and destinations, lower cost to serve first- and last-mile connections (for their rail lines), and faster trip times for many travelers.[30] It notes that in much of the sprawling Los Angeles region, this new microtransit model might be able to more effectively serve travel demand than fixed-route services and could help boost ridership on fixed routes by not only increasing trips by existing users but also attracting new users. The agency could increase the frequency of service on core routes and reduce or eliminate fixed-route service on "low-performing" routes. And it could use the microtransit services for underserved areas and to enhance paratransit offerings.

A big challenge for LA Metro and other legacy transit operators, all owned and operated by cities and regional governments, is how to comply with a stream of requirements that impose costs and limit flexibility. In its proposal, LA Metro says that a contracted microtransit operator must utilize union labor for operations and maintenance, as well as comply with Americans with Disabilities Act rules, Title VI rules prohibiting discrimination, environmental justice regulations, all applicable Federal Transit Administration (FTA) policy, and federal, state, and local law—presumably including "buy America" requirements.

Handing Off Demand-Responsive Services

Another market opportunity for microtransit is paratransit, or demand-responsive transit, offered by legacy transit operators either directly or contractually. These curb-to-curb services, mainly targeted at those with physical disabilities, are federally mandated—and very expensive. They have seen the largest rise in cost of any transit mode since the late 1980s, even though they carry only about 2 percent of all transit trips. The average cost exceeded $23 per trip in 2013, but riders paid an average fare of only $2.61, barely 10 percent of the cost.[31]

Many of these paratransit trips could be served by Uber, Lyft, and other ridehailing services—or by microtransit. A 2016 Brookings Institution report[32] estimated that transit agencies could save $1.1 billion to $2.2 billion per year (based on an average cost of $13 to $18 per ride) by using ridehailing companies for paratransit. In Boston, the Massachusetts Bay Transportation Authority (MBTA) ran a six-month pilot program to allow patrons to take a taxi instead of its RIDE paratransit service, which costs taxpayers millions to operate and which critics say is frequently late and takes wildly inefficient routes. Advocates for the disabled urged studying the feasibility of adding Uber to that fleet, and the MBTA responded by starting a separate year-long program in September 2016 to subsidize Uber and Lyft rides for customers with disabilities.[33]

In October 2016, the FTA announced project selections for its Mobility-on-Demand Sandbox Program to examine how public agencies can integrate emerging technologies to improve demand-responsive service, reduce costs, and improve other service and operational efficiencies. A number of selected sandbox projects are testing the potential for shared mobility to provide on-demand paratransit-like services. Evaluation of these projects is anticipated to be completed in 2018–19. It remains to be seen whether paratransit innovations enabled by technology and new partnerships can overcome challenges such as the fact that lower-income and elderly people often lack smartphones.

Integrating Innovative Technologies

Just as with cars, the integration of innovative technologies provides opportunities to improve transit service and reduce costs. Vehicle automation attracts the majority of attention, but there are many other promising innovations, including in-vehicle camera security systems, remote vehicle operation monitoring, various driver-assist technologies and systems, ridehailing tools, and electrification.

Ridehailing Tools

Public transit could benefit from embracing techniques and technologies pioneered by ridehailing companies and emulating their service mentality. This means taking advantage of web- and smartphone-based apps to give travelers convenient access to information about travel options and provide real-time updates.[34] Such apps can provide information to travelers to help them plan a trip using different modes, even in unfamiliar places.

This effort is well under way and will no doubt spread. For example, OneBusAway is an open-source platform for real-time transit information developed by the University of Washington and currently available in Atlanta, Seattle, Tampa, York, New York City, and Rogue Valley.

The RideTap software development kit and network from Moovel (owned by Daimler) makes it easier for transit agencies to integrate their services with those of other shared mobility providers without having to develop their own applications from scratch. The first RideTap program was launched in Portland, Oregon, in May 2016 with Car2Go, Lyft, and TriMet, Portland's public transit agency.[35] The TriMet ticketing application now enables riders to book Lyft, Car2Go, and Biketown bikeshare services from a single app instead of toggling between multiple applications.

In 2015, Dallas Area Rapid Transit (DART) began to integrate Uber into its GoPass mobile app to help riders plan and pay for "complete trips."[36] While many jobs exist within a half mile of DART's services, the areas are not easily accessible on foot. The ability to access Uber through DART's GoPass app gives riders an option to get to those nearby locations and encourages others who had not considered using DART to do so in tandem with the ridehailing service. Similarly, Metropolitan Atlanta Rapid Transit Authority (MARTA) has integrated access to Google Transit Trip Planner and Uber into its MARTA On the Go app, which also gives real-time bus and rail information.[37]

Ridehailing tools could also improve paratransit. By making it possible to coordinate service across providers and agencies, technology could allow more companies to offer paratransit services. In this way, the unique needs of various travel markets could better be met. For example, many agencies operate large numbers of vehicles capable of handling wheelchairs and mobility aids. These vehicles are used even for travelers who don't need mobility aids and could be replaced by smaller, less expensive vehicles. Similarly, paratransit vehicles could meet the immediate needs of other riders if they were running in on-demand mode.[38] Custom-designed software could optimize routes for vehicles that riders could order from kiosks or desktop computers.

Electrification

As indicated earlier, buses in the United States on average have higher greenhouse gas emissions than cars per passenger mile. Light- and heavy-rail systems perform better, largely because they are electrified. Now buses can also be electrified. Electric buses have zero emissions, reduced vehicle noise, and lower operating and maintenance costs.

Today, transit buses are mostly powered by gasoline, diesel, and natural gas, but batteries are starting to become competitive.[39] China has taken a leadership role in electric bus manufacturing and demand, with electric bus sales exploding a hundredfold in five years, from 1,136 in 2011 to 115,700 in 2016, grabbing a 20 percent market share.[40] In the United States, battery electric buses are progressing steadily but more slowly. They are being purchased by transit agencies from Seattle to Tallahassee, with some agencies making commitments to transition their entire fleets to electric propulsion. Companies like Proterra, BYD, Ebus, New Flyer, GreenPower Motor Company, and Mercedes-Benz with its Citaro E-CELL are competing for the contracts. Dramatic improvements in battery cost and capability are accelerating progress.

Automation

The most tantalizing opportunity is automation. Automating transit buses and vans will allow a massive restructuring of public transportation services, in part because today's transit system is designed around the high cost of drivers. Automating some or all fixed-route and demand-responsive bus services will greatly reduce operating costs. Analysis of the National Transit Database indicates that bus drivers account for 42 percent of bus operating expenses. Transit operators typically utilize large buses, as opposed to vans and small buses, largely because they are trying to increase the productivity of (expensive) drivers by loading more passengers per vehicle. If relieved of driver costs, transit agencies could run smaller vehicles at higher frequencies and thereby provide the more frequent service desired by travelers.

Buses have attractions and advantages in deploying emerging automation technologies. As an arm of local governments, they have well-established operational protocols and insurance. With their high rate of utilization, they rapidly accumulate service time; a public transit bus commonly accumulates more than 35,000 miles and 2,500 hours annually. They have professional operator and maintenance staff, high public exposure, and established fleet facilities. They often operate in urban environments in close proximity with vehicles, pedestrians, and cyclists. They also operate on fixed routes, providing a defined physical environment for autonomous operation. All these factors make buses an attractive laboratory for early application of safety-enhancing vehicle automation technologies. The potential to improve performance and safety in these environments can help justify the policy, investment, and regulatory changes needed to test and deploy these new technologies more broadly.

Automation of connected buses will also enable transit agencies to manage operations better. They will be able to assess ridership in real time and dynamically adjust the bus supply to maximize ridership.[41] And because driverless buses will be able to follow one another more closely, the same roadways will accommodate many more travelers. A study at Princeton University showed that reducing the time between buses from five seconds to one second would allow self-driving buses to transport more than two hundred thousand passengers per hour through the Lincoln Tunnel on the 2.5-mile bus lane connecting the New Jersey Turnpike with Midtown Manhattan. Currently, fewer than forty-one thousand people are transported per hour through the tunnel on what is already the busiest bus lane in America.[42]

In addition, automation provides an opportunity to leverage technology to replace expensive infrastructure, a crucial shift. Until now, efforts to increase road capacity, including for transit, have focused on investments in rights-of-way and infrastructure. In the case of transit, such efforts have meant rails and exclusive lanes for bus rapid transit. Because of the high expense of such investments, high-speed transit has been restricted to a limited number of high-volume locations. In a fully automated and managed transportation network, public transit vehicles could be assigned to congestion-free mixed-vehicle lanes without depending on new guideway infrastructure or exclusive rights-of-way. If computers and sensors rather than tracks are guiding vehicles, and if technology rather than grade separation can be used to eliminate interactions, high-speed public transportation will cost substantially less than today's infrastructure-intensive solutions.

When will all this happen? Mercedes, for one, says it is investing roughly $227 million to develop semiautonomous city buses by 2020.[43] Driverless shuttles and buses operating on simple fixed routes are possible in the early 2020s, but it will probably be many years before driverless buses are broadly deployed.

How Can Policy Help Public Transportation Flourish?

Public transportation is on the cusp of dramatic change. New technologies and service models will have profound impacts on transit. Those impacts are inevitable—and likely transformational. Current ways of delivering and managing transit must and will change. More traveler choice and better service are likely—the heaven scenario of chapter 1—but not assured. A multiplicity of stakeholders, limited funding streams for transit, the needs of carless travelers, and the economic vitality and livability of cities frame the challenges.

Policy plays a crucial role—more than with any other transport mode. That is because transit is so heavily subsidized and because it fulfills so many public service functions. The impacts on mobility and quality of life are so significant that the consequences cannot be left solely to the pace of technology evolution, business strategies, and market adaptation.

Some things are clear. First, financial support must be adequate to sustain transit infrastructure and services in high-volume locations where trains and buses are uniquely suited. Second, part of the social contract for public transportation is that government will provide a safety net of affordable mobility for low-income urban travelers and ensure door-to-door assisted services for travelers with physical limitations. To the extent that alternative mobility options undermine the provision of public transportation as we know it, remedies will be needed.

Third, as private companies begin to play a larger role, oversight will be needed. Local governments should not stifle innovation, but at the same time, they must protect the public interest. They must ensure competition among private-sector providers while monitoring for abusive practices and ensuring equitable access, including during disasters. They must guarantee that those unable to afford or use apps and debit/credit payment systems are served through either a third-party interface or an alternative means of trip scheduling and fare payment. Ideally, they will create a universal portal for access to the various mobility options.

Fourth, transit will have to address labor implications. The National Transit Database reports approximately two hundred thousand bus-operating employees for fixed-route services and an additional one hundred thousand employees for demand-responsive services, with more than one thousand agencies providing bus-transit services. In the near term, demand for drivers will likely increase as the use of vans and small buses increases. While driverless buses are still far off, policy makers will need to address the eventual automation of vehicles. It will be important for the public and policy makers to recognize that a driver's functions go beyond guiding the vehicle and include oversight of fare collection and customer behavior while in the vehicle and providing customer information. Who provides these functions if the vehicle is automated?

Fifth, one of the most critical issues facing transit stakeholders, as well as the broader transportation planning community, relates to long-range planning and capital investment. The large cost of many fixed-infrastructure commitments and the fact that these assets have very long lives require timely consideration of how to proceed. For example, rail projects that agencies are committing to today might be inaugurated just when new automated vehicles (AVs) appear, potentially cannibalizing their markets.[44] In addition to providing direct competition, new mobility options could influence development patterns, further impacting travel demand. Public transportation stakeholders have to address this challenge in a responsible fashion to retain credibility with the public as stewards of public resources. Future corridor plans should evaluate automated transit options and consider AV services as they evaluate investments to meet future mobility needs. Potential strategies include robust scenario analysis, preferential treatment of vehicles that carry large numbers of passengers on managed lanes, and disciplined commitments to new infrastructure.

The path forward requires tearing down barriers between transport modes, perhaps more quickly and deliberately than ever before. At the extreme, this means affected groups—users, local governments, taxpayers, operators, advocates—reorganizing around mobility or quality-of-life objectives, rather than around modes or technologies. Instead of being wedded to building a new rail line or subsidizing a new bus line, interest groups could focus on serving low-income riders or reducing pollution.[45] Progress will require leveraging and directing the entrepreneurial private sector in such a way that it can complement the purposes that have sustained the historic public investment in transit.

Key Policy Strategies

The aim is to enhance public transportation to expand accessibility, improve urban livability, and protect the environment while continuing to provide a social safety net for those less advantaged.

At the National Level
- Reform federal public transit and highway finance to support all urban pooling services, including microtransit companies and ridehailing companies carrying multiple riders, while guaranteeing base funding for legacy transit operators.
- Restructure transportation formulas used to allocate subsidies to focus on customer outcomes and to be agnostic about service providers.
- Provide public funding for the purchase of battery and hydrogen fuel cell electric buses and for increasing the supply of electricity and hydrogen to pooling, microtransit, and legacy bus services.
- Employ web- and smartphone-based apps to give travelers convenient access to real-time information and travel options, facilitating travel planning with different modes, even in unfamiliar places.
- Use third-party data brokers such as universities to make it easy to aggregate and analyze privately generated data to enable better service planning without compromising intellectual property or competitiveness.

At the State or Local Level

- Reform state and local public transit subsidies to support all urban pooling services, including microtransit, ridehailing, and taxi companies carrying multiple riders, while guaranteeing base funding for legacy transit operators.
- Provide technical support to public transit operators and cities to partner with private app-based services to provide first- and last-mile services.
- Continue mandating and financially supporting services by transit operators or others to meet the needs of low-income communities, low-income travelers, and those unable to drive.
- Enact congestion pricing on roads and use revenue to support the entire suite of transit and pooling services.
- Redefine public transit agencies to serve as a centralized portal for mobility, providing trip planning, trip scheduling, and revenue collection services—in parallel with commercial businesses.
- Integrate information and payment for services across public transit, ridehailing, bikesharing, and carsharing to streamline access for all travelers, rich and poor.
- Fund programs to transition riders who are physically able from paratransit services to app-based, demand-responsive services.
- Contract with microtransit companies to use vans and small buses to serve locations with little or no conventional transit service.
- Encourage microtransit and first- and last-mile services by providing access to curb space, parking, and rights-of-way.
- Require ridehailing services to allow prepaid debit cards (such as the Chicago Ventra card) for those without credit cards.

CHAPTER 6

Bridging the Gap between Mobility Haves and Have-Nots

Anne Brown and Brian D. Taylor

The three revolutions collectively offer considerable promise to increase mobility for the poor, the elderly, people with disabilities, youth, and other historically disadvantaged communities, but policies are necessary to ensure this hoped-for outcome.

GRACE IS A SINGLE MOM WITH TWO kids living in Koreatown in Los Angeles. High housing costs have put car ownership out of reach for Grace, so she regularly suffers through a long, complicated morning and afternoon travel grind. Each weekday, she rises at 5:30 a.m. to dress and feed her children and walk them four blocks to her cousin Lydia's apartment; Lydia then walks Grace's daughter to daycare and her son to elementary school while Grace makes a seventy-five-minute, two-bus trek from Koreatown to her job as a teacher's aide in Westchester. The trip home in the afternoon is just as lengthy and complex, and

Grace struggles to get dinner on the table for her children by 7:00 p.m. each evening.

Unlike Grace, those with access to cars generally enjoy high levels of mobility. They travel when and where they want and usually park for free. Some transit users, especially those in dense city centers, enjoy good access to jobs, school, health care, and education. But many other travelers do not, especially those without cars—as illustrated by the fictional Grace. Outside the United States and other rich countries, even more suffer from limited mobility and access. With the advent of new mobility services, like Lyft and Uber, those with smartphones and credit cards have an expanding array of travel choices. But travelers without these tools are increasingly left behind. They face a vicious cycle in which limited access leads to lost educational and income opportunities, which leads to further limits on access.

These disparities between the mobility haves and have-nots vary systematically by income, race, ethnicity, physical ability, and location, all of which raise troubling questions. Public policies to address these mobility inequities have met with only limited success and have often not been a high priority for public officials. This is at least in part because of uncertainty and occasional disagreement about the best ways to increase mobility equity, particularly for low-income households.

Against this backdrop, the three revolutions—vehicle electrification, shared mobility, and vehicle automation—offer both the *promise* of expanding mobility and accessibility (the dream scenario) and the *peril* of exacerbating inequities (the nightmare scenario). Grace's daily three-plus hours of travel could get even worse if public transit is cut back because of the success of personal automated vehicles (AVs) and shared mobility services. Or her commute could get easier if policy choices put automated shared-ride electric vehicles (EVs) at her disposal. We are, in other words, at a crossroads where new mobility systems could be a plus or a minus in addressing the accessibility gap and, in turn, the opportunity gap.

We begin this chapter by taking a closer look at the current mobility inequities among different traveler groups in the United States, though the issues raised are not unique to the American context. We then consider how the three revolutions might be steered by policy to shrink rather than expand the mobility gap.

The Mobility Have-Nots: Dependent on Transit and Taxis

Much of the mobility divide between groups of travelers in the United States exists because of financial barriers to car ownership. Those who cannot afford to own cars have limited mobility options, particularly because until relatively recently, taxicabs have been the only way to gain car access without car ownership. For many Americans, living without a car is nearly impossible due to inadequate transit service and diffuse activities in suburbs, smaller cities, and rural areas.[1] Even the poor—who comprise the majority of public transit riders—travel primarily by car,[2] though access to cars and car reliability is far more problematic for the less affluent.

The importance of car access in the United States is reflected by the near ubiquity of auto ownership; the most recent data available show that in 2009, only about 8 percent of American households did not own a car. This statistic, however, disguises the dramatically uneven distribution of carless households within the population. While less than 3 percent of all households earning $50,000 or more per year did not own a car, more than one-fifth (21 percent) of households earning less than $25,000 did not own a car (see figure 6.1).[3] The divisions were equally stark by race and ethnicity; only 6 percent of white households did not own a car, while 22 percent of African American and 12 percent of Hispanic households were carless.[4] Unequal access to automobility manifests in fewer trips and miles traveled by low-income compared to high-income households.[5]

While living without a car can impede mobility, private vehicle ownership comes at a high cost: the average sedan bought new costs about

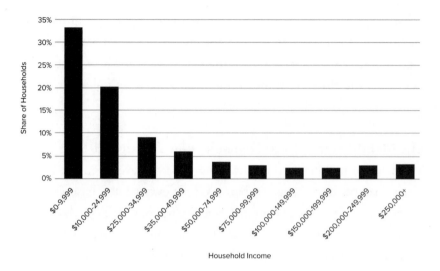

Figure 6.1. Share of households owning zero vehicles by income.

$8,500 to own and operate per year, and the average sedan bought used costs about $6,600 per year.[6] The average American household that earns $30,000 to $50,000 annually spends 17 percent of pretax income on car payments, gas, and maintenance.[7] But even owning a car does not always equate to car access, as both vehicle reliability and competition among household members in lower-income households—where licensed drivers often outnumber motor vehicles—can substantially reduce car access for many travelers. Low-income households tend to own both older and fewer cars per adult compared to higher-income households; and African American, Asian, and Hispanic households generally have fewer cars per adult than non-Hispanic white households.[8]

Together, competition among household members for access, ownership of older vehicles, and frequently high maintenance costs undermine reliable vehicle access in low-income households; only 63 percent of low-income households report having access to an operational car versus 94 percent of higher-income households.[9] In addition, for many households, car access is not stable over time. Poor households, and African

American and Hispanic households, tend to switch into and out of car ownership over time; car ownership is much more stable for wealthier, white, and Asian households.[10]

Carless travelers can, of course, get around using other means, such as carpooling, borrowing a car from friends or relatives, or using taxis, but these options tend to be less reliable and/or less affordable. Carpooling or borrowing a car is often logistically complicated, difficult, unreliable, or all three.[11] Taxis, to a degree surprising to many, have partially filled the mobility gap faced by carless households and are disproportionately used by very-low-income and carless travelers. For example, while carless households made only 4 percent of all the US person trips in 2009, they took more than half (53 percent) of all taxi trips. Similarly, households earning less than $25,000 per year made 17 percent of all person trips but 41 percent of all taxi trips.[12] Age and disability also influence taxi use; those with difficulty walking report increased reliance on taxis.[13] Adults over age sixty-five are increasingly reliant on cars, so some transit agencies and cities provide taxi coupons or vouchers to elderly and disabled travelers.[14] But taxis have historically been a high-cost option for carless travelers, and service has not always been reliably available in many minority neighborhoods, small cities, and rural areas.

Additional Challenges by Geographic Location
Indeed, not all places are created equal in terms of accessibility. Traditional fixed-route, fixed-schedule public transit works best in—and primarily serves—the dense centers of American cities. Residents of suburban and rural neighborhoods without regular access to cars are frequently mobility have-nots. Such neighborhoods are less dense than urban neighborhoods, typically requiring people to traverse greater distances in order to reach destinations. Increased distances present impediments to transit use, biking, and walking—particularly if direct, welcoming sidewalks are lacking and if large parking lots separate streets from destinations.

Policies intended to increase mobility in these areas have been relatively limited in both rollout and efficacy. Skeletal, circuitous transit routes with infrequent and sometimes unreliable service reduce the number and viability of alternative mobility options.

Limited alternatives to car-based mobility in suburban and rural areas are increasingly problematic as the general population, jobs, and the poor continue to suburbanize. Despite the revitalization of some large cities around the United States, half of all Americans lived in the suburbs in 2010, and most suburban populations continue to grow.[15] The poor too are suburbanizing rapidly. In 2010, the number of poor in the suburbs (about twelve million) eclipsed the number of poor in primary cities (ten million) for the first time, and between 2000 and 2008, the number of poor grew faster in the suburbs than in primary cities, smaller metropolitan areas, or nonmetro areas.[16] The rapidly growing elderly population will also be increasingly suburban in the future, with 90 percent of adults age forty-five or older reporting a desire to age in place.[17] Because of limited alternative transportation options in the suburbs, aging adults there who can no longer drive will face increasing mobility restrictions.

But even where fixed-route public transit works best—in larger, densely developed cities—it still provides inferior access compared to cars. In Detroit, cars provide better access to jobs because transit service can be unreliable or infrequent even in the most central neighborhoods.[18] In transit-rich San Francisco, only a limited number of neighborhoods enjoy high job accessibility by transit, while job accessibility by car is nearly ubiquitous across the region.[19] And consider the challenges for low-income travelers, who are more likely than higher-income travelers to work varied schedules and less likely to work a nine-to-five job.[20] Traditional fixed-route, fixed-schedule public transit service is most frequent during the peak hours and wanes during off-peak and late evening periods, sometimes dramatically. As a result, the commute schedules of workers in lower-paid occupations are less likely to align with the days and times when transit runs most frequently. Lawyers and accountants working regular business hours in city centers enjoy frequent inbound transit service in the mornings and frequent outbound service in the afternoons, while workers who clean those buildings each evening are likely to see less frequent service inbound in the afternoon and perhaps no service at all outbound late in the evening.

It's clear that the mobility challenges faced by the poor, the physically impaired, and residents of suburban and rural areas aren't being met and will worsen as the population ages and where land use is sprawling. The three revolutions offer a chance to change the cost and access calculus of auto-mobility and pursue true mobility equity. In particular, the rise of afford-able, on-demand shared mobility, plus automation, offers the potential to expand transportation options for carless people. On the other hand, the electrification and automation revolutions could unfold in ways that leave disadvantaged travelers worse off.

Electric Vehicles: Extending the Benefits to the Have-Nots

As noted earlier, most low-income people get around in cars, but those cars are almost always used and sometimes quite old. Vehicle innovations—whether seat belts, antilock brakes, or catalytic converters—are slow to filter down to low-income drivers. Electric propulsion is no exception. The first mass-market EV, the Nissan Leaf, came onto the market in 2010. Starting in about 2015, the used electric car market became a great place to get a deal, thanks to depreciation being far higher than for other car segments. For example, a five-year-old Leaf was selling for about 15 percent of its original sales price.[21] But consumers in the used market are less inclined to buy electric cars (see figure 6.2), in part because vehicle charging can present a stumbling block for individuals not owning a house, which is the case for most low-income used-car buyers.

Most EV owners depend on overnight charging at home and only secondarily on public and worksite charging.[22] Lower-income travelers

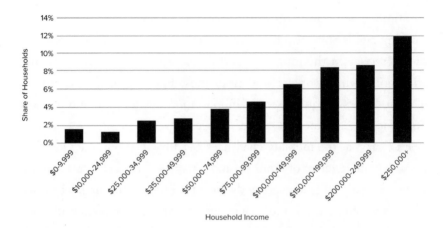

Figure 6.2. Share of households owning at least one hybrid/EV by income. Sample: All California households that reported income and own at least one car. Data source: 2012 California Household Travel Survey, weighted.

are much more likely than those with higher incomes to reside in multiunit dwellings with shared (or no) off-street parking, which makes home charging difficult or impossible. Homeowners with garages can easily add a home charging station, but few landlords in multiunit apartment buildings are willing to install chargers.[23] This lack of easy charging access might put EVs, even used ones, effectively out of reach for lower-income car buyers until and unless many more public charging stations are built. Unfortunately, the revenue from charging stations is so low that someone needs to subsidize them—the government, landlords, or employers. That is not a recipe for massive investment in public chargers.

Subsidies and rebates for the purchase of EVs have also been out of reach for mobility have-nots. The federal government and a variety of states offer EV purchase incentives, including rebates, tax credits, sales tax exemptions and reductions, and fee exemptions and reductions. But because these have almost exclusively been confined to the purchase of *new* vehicles, most of the benefits of these subsidy programs have flowed to higher-income households, which purchase the vast majority of new cars and almost all new EVs.[24] Studies have found such policies to be not only inequitable with respect to income but also less effective among those with higher incomes, because higher-income new car purchasers tend to be less sensitive to rebate savings than lower-income used car purchasers would be. The majority of early EV buyers would have purchased an EV without the public subsidy.[25]

Given the high barriers to electric car adoption among the mobility have-nots, what can be done to include disadvantaged travelers in the transition to EVs? Two sets of policies are being considered.

The fact that most low-income households reside in multiunit dwellings, where EV chargers are sparse, is part of a chicken-and-egg challenge whereby individuals are reluctant to buy vehicles without easy access to a charger and landlords are reluctant to invest in chargers if there is no demand. Four different measures might be taken to address this conundrum. First, induce or subsidize landlords to install chargers. Second, update building codes for new apartment construction to require wiring suited to installation of chargers (and perhaps require installation of chargers as well). Third, subsidize the installation of chargers in areas with large shares of multiunit dwellings. And fourth, educate low-income residents on the existence and potential benefits of EV rebate and subsidy programs.[26]

At the same time, vehicle purchase rebates and subsidies need to be restructured to assist lower-income buyers, who purchase used vehicles. California is experimenting with tiered rebate policies that can be used to direct the largest benefits to the poorest households, as well as policies that speed up turnover of the fleet's oldest and most polluting cars.[27]

Shared Mobility and Automation: Bridging the Divide

Electrification will probably make the biggest difference to disadvantaged travelers when used in fleets of shared vehicles. The same can be said of automation. Shared mobility services have the potential to expand automobility to those without easy access to cars, at lower cost and with greater reliability. These services—including bikesharing, carsharing, and ridehailing—can provide on-demand, point-to-point transportation at low cost.[28] Carsharing services like Zipcar and Car2Go offer a vehicle for those special situations when a car or a pickup or an SUV is really needed. In the past, some cities have partnered with federal agencies to waive carshare membership fees and reduce hourly rates for low-income or welfare-to-work households.[29]

California is currently experimenting with a low-income electric car-sharing program. The California Air Resources Board made a $1.6 million grant to Los Angeles in 2015 to establish a carsharing program in low-income neighborhoods and install charging stations to support the program.[30] The hope is that the program will take old, high-emissions cars off the city's streets while at the same time leaving families with more disposable income by eliminating car payments, fuel costs, and insurance and maintenance expenses. The program involved community groups in shaping on-the-ground details, since local knowledge will play a key role in the program's success. For instance, not everyone in the target area can be assumed to have a smartphone, a credit card, or an Internet connection, so a multilingual call center might be essential to help people reserve cars.

Early evidence suggests that ridehailing is already providing mobility to low-income and carless travelers. One study of ridehailing service users in San Francisco found that nearly one-third earned *under* the city median income and 43 percent did not own a car.[31] These ridehailing users drove less but took more trips than previously, suggesting that ridehailing increased their overall mobility. Services like Lyft and Uber also expand the geography of car access beyond that of traditional taxi services. In New York, they serve the city's outer boroughs better than taxis.[32]

As less expensive pooling services—such as Lyft Line, UberPool, and microtransit—become more widespread, low-income and physically disadvantaged travelers are likely to benefit disproportionately. And as these services expand, the cost per trip will be even less. Such services make door-to-door mobility more affordable for lower-income travelers. They thrive in dense urban environments, where the number of travelers is great and trips are short, but they are also likely to be attractive in suburbs and rural areas, where public transit is scarce or nonexistent.

Even lower costs will prevail when AVs become available through mobility service providers, thereby greatly expanding access for disadvantaged travelers. Because they can offer point-to-point services, AVs can extend automobility to those too young, too old, or too physically impaired to drive. The cost of such services is expected to be well below today's Lyft- and Uber-like services, as indicated in figure 1.2, since fully automated vehicles will save money by not requiring a driver.

Similarly, automation of transit—particularly of buses—could lower fares by greatly reducing labor costs, which are a substantial part of transit costs. While virtually all public transit service is subsidized and already priced well below actual costs, transit agencies could reinvest driver salary savings into reduced or eliminated fares or more extensive and frequent service. Automation might also help improve on-time performance and increase speeds by avoiding delays caused by sick or late drivers, transitions between drivers, and driver breaks. And better connectivity among buses and with traffic signals would allow buses (and perhaps vehicles with multiple riders) to travel faster and reduce delays. Increased transit reliability and faster speeds would enhance mobility for disadvantaged populations, who disproportionately patronize public transit.

Factors That Could Widen the Mobility Divide

Without supportive policies and programs, though, automated and shared services are unlikely to solve mobility problems for those without easy access to a car. This increase in the mobility divide could occur in three primary ways: lack of a smartphone and/or credit card, discrimination, and transit displacement.

Lack of a Smartphone and/or Credit Card

A person typically needs both smartphone access and a credit/debit card to use shared mobility services. Some people, especially those with lower incomes, do not have bank accounts or smartphones.

Only 7 percent of all US households were without a checking or savings account in 2015, but this percentage was higher among African American and Hispanic households and households with lower income, less education, younger members, and working-age disabled members.[33] Both Uber and Lyft require payment by credit or debit card.

Smartphone ownership is also problematic. In 2015, about one-third of all Americans did not own a smartphone—though only 15 percent of youth aged eighteen to twenty-nine did not. Overall smartphone ownership was higher among urban (68 percent) and suburban (66 percent) residents compared to rural residents (52 percent).[34] Smartphone ownership increased with education and income but was actually higher among African Americans (70 percent) and Hispanics (71 percent) compared to whites (60 percent), probably because the percentage of African Americans and Hispanics living in urban areas is higher than that of whites. While smartphone ownership continues to rise, a large number of people are currently excluded from using shared mobility services because they don't have a smartphone.

A policy solution to these credit and smartphone handicaps at the regional level is to integrate information and payment for services across modes (such as transit, ridehailing, bikesharing, and carsharing) to streamline access for all travelers, rich and poor. Fare-payment systems can be structured to offer subsidies to users of multimodal travel. Policies to allow those without credit card accounts to use prepaid debit cards can increase shared mobility access among unbanked populations, as demonstrated by the Chicago Ventra card.

Chicago's Ventra Card

Even though cash fares slow down transit boarding and electronic payments have become commonplace, most public transit and taxicabs continue to accept cash fares because cash remains an important fare medium for many riders, particularly unbanked travelers. By contrast, most new shared mobility services—including ridehailing and many bikesharing and carsharing systems—do not accept cash but instead require a debit or credit card to rent or hail a vehicle. This requirement might exclude the unbanked, who are more likely to be immigrant, undocumented, and less educated than the general population.[35]

Some transit agencies have moved to integrate transit cards with shared mobility options. For example, Los Angeles Metro's Transit Access Pass (TAP) smartcard can be used to rent bikes from three local bikeshare networks in Los Angeles, Beverly Hills, and Santa Monica, but users must still have a credit card to create a bikeshare account. Although the fare card streamlines access across modes for those with the requisite debit or credit card accounts, it does nothing to increase access to shared mobility for those without debit and credit cards.

The Ventra card program of the Chicago Transit Authority (CTA), introduced in 2013, is an example of one way that transit agencies can partner with shared mobility and also extend access to unbanked populations. Each Ventra card has two accounts: a transit account and a prepaid debit account. The transit account allows for traditional transit transactions, such as loading fares and passes onto the card for use on CTA buses and trains. The preloaded debit account allows purchases anywhere debit cards are accepted. The Ventra card does not charge interest or require a minimum balance. In addition, funds stored on Ventra cards are insured and can be recovered or transferred to another card if a Ventra card is lost or stolen.[36] The Ventra card offers new banking opportunities for currently unbanked populations and, in addition, extends mobility options to the unbanked, who can now use the Ventra card to access shared mobility services that require a debit or credit card. For example, the Ventra debit card number can be linked to a bikeshare or ridehailing account.

The Chicago Ventra card bridges the debit/credit card gap currently faced by unbanked residents that prevents them from accessing shared mobility services. By providing a transit card with debit card capabilities, the Ventra card does more than provide transit access: it opens up new doors to ridehailing, bikesharing, and carsharing, which were once out of reach for those without bank cards. The Ventra card offers an enlightening example of how public transit agencies can lead efforts to extend residents' mobility across shared modes.

Discrimination

A second way the mobility divide might increase in the coming era of shared mobility is through discrimination against low-income neighborhoods. For example, carsharing networks have been found to have a limited presence in low-income neighborhoods,[37] and evidence from London suggests that lower bikesharing usage by low-income travelers likely results from docking stations located outside their neighborhoods.[38] A 2016 study of more than 1,500 ridehailing trips found that Uber and Lyft drivers were more than twice as likely to cancel rides requested by African American patrons. Gender bias might also be present; men traveling in low-density areas were more likely than women to have their requested trips canceled by the driver.[39]

In the short run, city and regional planners can take steps to promote more equal access to shared travel modes. For both bikesharing and carsharing networks, planners and civic officials must consider policies to overcome financial and physical barriers to accessing these new and expanding services. Perhaps the best way to address the ridehailing discrimination divide is for local and regional government officials to obtain access to private ridehailing company data to ensure geographically equitable service coverage that does not discriminate by race, ethnicity, gender, socioeconomic status, or disability. One way to obtain such access would be for cities to give permission for mobility providers to operate only on the condition that such data are shared.

Expanding Bikesharing to Serve Disabled Travelers

Bikesharing programs have made concerted efforts to extend services to some mobility-disadvantaged travelers, such as low-income communities and riders, but less so for others such as the disabled. While it might seem that physical limitations would preclude the disabled from bicycling altogether, many disabled travelers can use adapted or accessible bikes such as handcycles, quadricycles, and tricycles. In London, for example, 70 percent of disabled adults report being able to cycle, but only 6 percent regularly or occasionally do.[40] This suggests a disconnect between physical ability and supportive infrastructure. While adapted bicycles have flourished in the private bicycle market, they continue to be largely absent from bikesharing systems.

Urban and transportation planners, while often cognizant of geographic challenges to including low-income communities in bikesharing networks, rarely explicitly include disabled travelers in plans for bikesharing. Given current alternatives to traditional bikes, however, this need not be the case. To improve disabled residents' access to shared mobility opportunities, cities can provide accessible bicycles within bikesharing fleets. Accessible bikes can easily be introduced in bikesharing systems that secure bikes using mounted U-locks, without requiring additional or retrofitted infrastructure. In bikesharing systems that use fixed docking stations, new docks will need to be added to accommodate wider accessible bikes. While adding this capability will add cost, shared mobility access can extend mobility options for all residents.

Although the overwhelming majority of bikesharing systems do little to include disabled riders, four small US cities currently operate fully accessible bikesharing programs that offer adapted bikes: College Park, Maryland; Carmel, Indiana; Corvallis, Oregon; and Westminster, Colorado.[41] The University of Ohio also has an adaptive bikesharing program. The systems are all operated by the bikesharing company Zagster—the largest and fastest growing bikesharing provider in the United States—and are marketed as inclusive bikesharing programs. Biketown, the bikesharing program in Portland, Oregon, introduced accessible bikes in July 2017, making it the first large-scale bikesharing system accessible to disabled residents.

These programs yield insights into how other bikesharing systems can expand shared mobility access to disabled travelers, though five of the six identified here are limited in scale and have relatively few adaptable bikes. As a short-term solution, mBike in College Park hopes to better connect disabled travelers to existing adaptable bikes by allowing users to search for adaptable bikes on its bikesharing phone app.[42]

Transit Displacement

A third way that shared mobility and automated services might exacerbate mobility inequity is by undermining fixed-route, fixed-schedule public transit and taxi services. This can happen where shared mobility providers draw customers away from these legacy services. In some instances, transit managers might respond by shrinking the number of routes or the frequency of service, and taxi operators might respond by reducing fleets and/or the time span of service. In other places, shared mobility might replace legacy services altogether.

Public transit agencies can anticipate and head off these threats to affordable mobility and turn adversity into opportunity. As suggested in chapter 5, they can enter into contractual relationships with Lyft, Uber, and others to extend the effective reach of trunk bus and rail lines through seamless, affordable first- and last-mile services and eliminate cost-ineffective routes (often in outlying areas) and expensive services, such as late-night operation. Such steps can increase public transit patronage overall and promote improved and expanded service in the process.

While supplementing or replacing transit with ridehailing services might provide more accessibility to more people, as well as increase transit efficiency, it also has the potential to make some travelers worse off if they cannot access these transit alternatives. The risk of reduced service to elderly, wheelchair-bound, or sight-impaired travelers is exacerbated with AVs. Such travelers might require assistance from their home to the AV and from the vehicle to the destination door; young children might require supervision in the vehicle as well. For such mobility-impaired travelers, eliminating the driver will not necessarily result in increased mobility. To ensure that displacement of transit services does not hamper their mobility, concerted efforts are needed to continue specialized transportation services for the elderly and the disabled.

Cities should ensure that taxi and ridehailing services—including wheelchair-accessible vehicles (WAVs) and trained drivers—are available to these populations, but requiring *all* vehicles to accommodate disabled passengers or wheelchairs would not be effective or efficient. Instead, cities are experimenting with service quantity and quality standards to provide reliable service for elderly and disabled passengers. For example, the City of Ottawa requires that 15 percent of service hours be provided by accessible vehicles, and Portland, Oregon, requires that WAV wait times be no more than ten minutes longer than non-WAV wait times. Ridehailing companies and taxis meet these goals by either providing accessible services themselves or contracting with paratransit services. Many cities (including Ottawa, Portland, Seattle, Toronto, and Austin) also levy fees on every taxi and ridehailing trip (typically $0.10 to $0.25) to fund improved accessibility services.[43]

More Policy Support for Mobility Have-Nots

Besides addressing the equity-related challenges that might derail potential gains from shared, automated, and electric transport, public policies can help reduce costs for disadvantaged travelers in other creative ways. For example, policies can encourage public-private partnerships among transit agencies, ridehailing services, and carsharing and bikesharing operators to create multimodal transportation hubs in low-income communities.

Policy makers can also encourage the development and deployment of cost-comparison tools or apps, perhaps even requiring service providers to furnish data that allow for real-time, apples-to-apples comparisons. RideScout, which originated in Austin, Texas, in 2011 and was acquired by Daimler in 2014, was just this sort of app; it was a popular tool before it was discontinued as obsolete in the summer of 2016.[44] Daimler's Moovel is a platform that allows comparison of various mobility services all over Germany. Transit App and Citymapper let riders compare transit options in selected cities around the world.

The benefits of such apps or tools are twofold. First, they increase transparency across modes, allowing travelers to select a mode—be it ridehailing, carsharing, transit, or bikesharing—to minimize time or cost for trips based on personal needs and preferences. At the same time, cost-comparison apps or tools—similar to websites that compare costs across airlines—encourage transportation services to be cost competitive, especially benefiting low-income travelers.

The Time Is Now

The three transportation revolutions will bring many benefits, but not necessarily to everyone. Of special concern are those who are already disadvantaged because of income, age, location, and/or physical capabilities. Many could be further disadvantaged by the three revolutions, depending on how new services and technologies are rolled out and how policy makers respond to them. In this chapter, we have examined equity challenges and opportunities for policy makers at the local, regional, state, and federal levels. For each, we have considered who should devise and execute policies in the short, medium, and long term to steer the three revolutions toward more equitable mobility. Indeed, a unique opportunity exists to reboot access and mobility services to *increase* mobility for disadvantaged segments of the population, but such mobility increases will require supportive policies to be fully realized.

The wheels of government can move slowly. Today, most regional transportation plans do not mention automation and rarely acknowledge the shared mobility revolution already under way, although local and regional planners are increasingly paying attention to accelerating developments. A few of the largest US metropolitan planning organizations, including those in Atlanta, Philadelphia, San Francisco, and Seattle, are beginning to take steps toward an increasingly automated future by developing planning scenarios, analyzing travel behavior, and considering changes to regional transportation infrastructure and services.[45] Such efforts need to explicitly consider the effects of automation, vehicle electrification, and shared mobility on the mobility disadvantaged.

For those without access to cars, perhaps the most promising development is the rise of pooling services—such as Lyft Line, UberPool, and microtransit. Indeed, because of the lower costs of these services, low-income and physically disadvantaged travelers are likely to benefit disproportionately, purchasing their automobility one shared trip at a time. This mobility could be further expanded by extending transit subsidies to disadvantaged riders who use these new sharing services.

These are not issues to be set aside for later; experience tells us that the time for policy innovation and experimentation is in the midst of transition, before stakeholder positions harden and change becomes more difficult. Without early policy interventions, the mobility gap between the haves and have-nots might well widen into a chasm. To ensure that Grace will someday find it easier and more affordable to get from her Koreatown home to her job in Westchester, the time to start planning for this future is now.

Key Policy Strategies

The aim is to advance social equity with shared, automated EVs by reconfiguring the public transportation safety net and incentivizing low-cost mobility services for mobility-disadvantaged communities.

At the National Level
- Fund demonstration projects that help those with limited incomes and in rural areas overcome obstacles to shared mobility and EV ownership.
- Provide (user-side) subsidies to the mobility disadvantaged, rather than subsidizing all travelers (with supply-side subsidies), in order to more cost-effectively serve the mobility disadvantaged.
- Ensure that mobility options for travelers with disabilities remain robust, especially as traditional transit and taxi services are reduced.

At the State or Local Level

- Expand efforts to include and engage disadvantaged communities in transportation planning, especially regarding shared mobility.
- Increase incentives and resources for EV charging infrastructure in low-income neighborhoods and rural communities.
- Require that mobility providers offer online or telephone booking, in addition to smartphone app-based booking, for those unable to afford or use smartphones.
- Require that providers offer alternatives to credit or debit card payment, such as prepay options, where passengers deposit money ahead of time in debit card–like accounts.
- Provide (user-side) subsidies to the mobility disadvantaged, rather than subsidizing all travelers (with supply-side subsidies), in order to more cost-effectively serve the mobility disadvantaged.
- Encourage public-private partnerships among transit agencies and ridehailing, carsharing, bikesharing, and microtransit providers to create multimodal transportation hubs in low-income communities.

CHAPTER 7

Remaking the Auto Industry

Levi Tillemann

The twentieth century was the age of automotive manufacturing.
The twenty-first century will be the age of mobility. Automakers will
remake themselves into companies that sell mobility services instead
of vehicles—or at least they will try to.

HENRY FORD BUILT HIS FIRST VEHICLE, WHICH he called the quadricycle, over the course of about six months. Both his wife, Clara, and his
friend James Bishop chipped in substantially. A decade and a half later, the
Ford Motor Company built its ten millionth automobile. The comparison
to Henry Ford's quadricycle effort is stark: the Model T factory employed
roughly five thousand times as many laborers and took about 1/2000th
the amount of time to build a car.[1] This staggering increase in scale and
efficiency was facilitated by the introduction of the assembly line in 1908.
Over the coming century, that simple innovation would expand the frontier of manufacturing efficiency to dizzying effect. The assembly line would
progressively destroy and reconstitute the global economy in a giant wave
of creative destruction—forever optimizing efficiency of production and

churning out goods for a world of hungry consumers.[2] Today, that line remains the beating heart of the global manufacturing economy.

But now Ford's mass production is about to be devoured by another technological revolution. To many, the name Elon Musk—and his corporate opus, Tesla Motors—epitomizes disruptive innovation in the twenty-first century. But a decidedly less heroic Silicon Valley CEO unleashed forces that will rival Henry Ford's assembly line in terms of economic impact: Uber's Travis Kalanick.

As Uber's CEO, Kalanick had a reputation as a ruthless competitor.[3] Unlike Musk, he was not beloved by the media or general public—quite the contrary. Part of this had to do with his cutthroat business practices; the fact that he did not produce shiny objects likely didn't help either. Kalanick was less interested in supercars or rockets than dollars and cents. But what he *did* create is a business model with a raw economic logic that makes SpaceX's rocket boosters and Tesla's electric supercars look puny by comparison—and for which Uber has been rewarded with an implied valuation that by 2017 exceeded that of Ford and GM, as well as Tesla.[4]

Uber's mobility-as-a-service business model—which, it turns out, Tesla is hard at work copying—might well lead to the demise of those automotive producers who fail to adapt. It might eventually end private vehicle ownership altogether and will almost certainly diminish the concept of the consumer of goods that has shaped our understanding of modern economics. The spillover effects will rival those of the Internet or the assembly line.

Today, Henry Ford's assembly line is still engaged in providing large volumes of things we buy, utilize, and discard—sometimes called stuff. But starting with capital-intensive goods like automobiles and moving on to other things we use only periodically—like sports equipment and tools—services are going to start to displace massive volumes of stuff in the economy of the twenty-first century. The foundation of this "Peers Inc." economy has been laid by Uber, Lyft, Airbnb, and eBay.[5]

The impacts of the mobility-as-a-service revolution will be replicated in industries throughout the global economy. The rise of service providers will lead to fundamental changes in motivation and incentives for owners of capital. It will allow for new business models that improve the efficiency of capital utilization and cross-pollinate in unforeseeable ways. Uber's approach to providing mobility as a service—rather than as an automotive product—typifies this transformation. In a very real sense, it exemplifies how mobility as a service will extend the long arc of corporate efficiency maximization beyond manufacturing and production and into the realm of consumption.

That much is settled. What remains unresolved is the question of how automotive manufacturers will respond.

From Producers and Consumers to Proprietors and Patrons

For a century, assembly lines disgorged wasting assets into the waiting arms of "consumers." The consumers' voracious appetite for goods paired with planned obsolescence fueled the global economy. This consumer gluttony was mirrored by a simultaneous quest for corporate efficiency in production.

It all started with Henry Ford's invention of mass production. After World War II, Toyota and Japan's industrial giants honed that efficiency through the tenets of lean production.[6] Today, Honda and others have instituted policies to eliminate waste from the manufacturing process entirely, and BMW has led the way in transforming manufacturing facilities into net-zero-emission operations.[7] It takes only thirteen to thirty-three person hours for a major manufacturer to build a car,[8] which sells on average for about $33,500 in the United States.[9] This rate of production is possible only because the industry is extensively mechanized and extremely efficient.

Although efficiency is generally extolled as a virtue, from the standpoint of resource consumption, there is something potentially toxic about an ecosystem that pairs efficient producers with ravenous consumers, with limited incentives for recycling and sustainable resource management. It leads, ultimately, to huge quantities of waste products in our air, on our land, and in our oceans.

Fortunately, this producer-consumer dyad is about to be challenged, and in some sectors replaced, by a new symbiosis—one that is more efficient and rational, traffics in services instead of stuff, and has a distinctly lighter environmental footprint. This new pairing consists of proprietors and patrons.

Massive, near-instantaneous, and virtually free exchange of information is allowing service providers and proprietors of capital to match up with patrons in ways not previously possible. This has led to new models of consumption and utilization of housing (think of Airbnb), automobiles (for example, Turo), and even clothing (sites like SnobSwap). In the future, these exchanges will be made even simpler because the service economy will increasingly depend on robots as opposed to people to provide services—reducing costs and eliminating much of the variation in quality of services delivered.

In many parts of the economy, ownership for individual use will become outmoded. Proprietors will own capital (much of which will consist of algorithms and robots), and patrons will purchase services provided by that capital stock. Proprietors and patrons will largely supplant producers and consumers.

Fleet Logic, Pooling, and Automation

Nothing exemplifies the power of this new business model as much as the automotive industry. In the United States, personal automobiles are utilized only about 5 percent of the time,[10] and on any given trip, the average car carries only 1.6 people—with 38 percent of trips carrying only the driver.[11] As mobility companies increasingly sell transportation not as a product (cars and trucks) but as a service, they will attack these inefficiencies with the same gusto Henry Ford and his successors (like Taiichi Ohno, the father of the lean Toyota production system) did manufacturing inefficiencies.

The single most powerful technology in this transformation is not going to be a better battery or a more fuel-efficient engine but a smartphone (providing geo-location and a consumer interface) and advanced routing algorithms (enabling on-demand pooling). The leap to driverless vehicles will accelerate this trend and supercharge the quest to supplant privately owned vehicles with heavily utilized, proprietor-owned cars that service many patrons. The ultimate result of this transformation will be fleets of connected automated vehicles (AVs) that reduce reliance on labor in transportation, much as mechanization and automation did in manufacturing and agriculture.

The fact that the product sold by successful automotive companies will not be a car but instead a service (mobility) will lead to a changed constellation of incentives. The new prevalence of ridehailing companies such as Lyft and Uber will have massive cascading effects on the automotive manufacturing industry and the broader economy. For one, the intensive utilization of vehicles by Uber, Lyft, and others will result in a much faster turnover of cars. Rather than being on the road 5 percent of the time, driverless fleets will be on the road 40 or 50 percent of the time. Rather than accumulating ten thousand to fifteen thousand miles a year as is now typical, they will accumulate one hundred thousand miles or more. Rather than lasting ten to fifteen years as is now common, they will last only two to three years. Faster turnover means that new technologies, including automation and electrification, will penetrate the market more quickly.

The intensive utilization of vehicles will also transform the types of vehicles purchased and used. Rather than carrying one passenger, they will carry three, four, five, or more. Indeed, many cars will be replaced by van-like vehicles operating as microtransit services that carry larger numbers of passengers. They won't be retrofitted cargo vehicles or min-ivans, as is current practice for vanpools, but designed for the comfort and convenience of passengers. The manufacturing industry will have a major new product line with a new passenger-centric design ethos.

And the intensive utilization of vehicles will transform passenger travel. It will have an exponentially greater impact than suggested by the number of cars managed by mobility service companies. These cars will serve as a platform for carpooling among strangers. One pooling vehicle might do the same work over the period of a day as twenty or more of today's individually owned vehicles.

Cost sensitivity to operational efficiencies means that proprietors will demand the most efficient technologies available from auto manufac-turers. Before long, electric vehicles (EVs) will take center stage, as they will eventually be cheaper to own and operate, especially when used intensively. This will reduce the number of vehicles required to fulfill transportation needs and the amount of oil they require to operate.

Fleet logic is what I call this cost-conscious approach to purchasing and operations.[12] Perhaps more than anything else, it is fleet logic that differentiates the incentives in the producer-consumer economy from those in the proprietor-patron economy. By allowing consumers to pur-chase services and allowing proprietors of capital to address operational efficiencies—and shoulder operational costs—the proprietor-patron eco-nomic model will put choices regarding fuel consumption into the hands of sophisticated corporate decision makers who have a much stronger incentive to conserve energy than an individual consumer.

Just look at airline fleets, which already operate on a proprietor-patron model. They feel the cost of fuel prices directly, so they are intensely focused on increasing efficiency in their operations. Fuel is responsi-ble for about 16 percent of their total operating expenditures; fuel plus labor accounts for around 48 percent.[13] Largely as a result, the airline industry has already seen intensive cockpit automation, is using larger planes (that is, more intensive pooling), and has reduced greenhouse gas emissions per passenger mile by roughly 50 percent since the 1970s—in the absence of fuel economy regulations.[14] In a fleet of vehicles, just as in a fleet of airplanes, corporate profits will depend on fleet logic driving more efficient utilization of proprietor-owned capital. Environmental progress will come along in toto.

Pooling will also increase transportation efficiency and improve environmental outcomes. While the emissions reductions in an EV are largely dependent on the source of electricity, the emissions reductions of a pooled ride are directly related to the number of riders. An EV might have an emissions profile comparable to a gas-guzzling twenty-miles-per-gallon SUV if powered by electricity from coal or might be essentially carbon-free if powered by electricity (or hydrogen) from renewable energy.[15] With pooling, each additional passenger adds minimally to the energy consumption but eliminates another vehicle from the road entirely—effectively doubling the efficiency of the vehicle with two riders, or tripling it with three people, or quadrupling it with four. This means that a pooled Prius is likely to be substantially less polluting than a nonpooled EV. The major goal of UberPool and similar services is not to save energy but to limit capital and operational costs, thus improving profitability. But if Uber can pack a lot more paying passengers into a single vehicle, even if powered by gasoline, the public benefits of carpooling are still large.

Automation will accelerate all these changes. Automation improves the economics of vehicles under almost all realistic scenarios—even in a hybrid system or transition phase where cars are partially automated and occasionally driven from a centralized control room. While fuel economy will eventually be important to proprietors, at first they will be even more interested in automation. Indeed, barring policy intervention, level-5 driverless cars will likely precede an intensive focus on fuel economy or electrification. That's because from a business-model standpoint, automation (elimination of the outsourced driver) is what brings labor costs down and brings capital and operational costs in house. (Today, all those costs are borne by the driver.) To recap, today Uber and Lyft rent two major assets: driver and car. They expect the driver to pay the cost of fuel—which is another major operating expenditure. But of those two cost centers, the driver is by far the more expensive. The outcome of these changes—the integration of pooling and automation—is vehicles being used more hours and carrying more people, at less cost, which will lead to ridership spiraling up.

In one real-life example, a San Francisco Uber driver recently informed me that he made $90,000 the previous year exclusively through driving Uber. That's more than the base cost of a Tesla Model S or Model X—not to speak of a cheaper Model 3, Toyota Prius, or Chevy Bolt. Even if a driver makes only a third that amount, that driver is a recurring expense—a high cost imposed every year. Further, that driver can drive only so many hours a day. Assuming an eight-hour workday five days a week, plus two weeks unpaid vacation, a driver making $30 per hour will take home approximately $60,000. An autonomous vehicle with no vacation and a thirteen-hour workday would perform work equivalent to about $101,000 of driver time in the same twelve-month period. Eliminating drivers will yield massive operational efficiencies and allow mobility companies to provide service at a much, much cheaper price.

By the simple laws of economics, a reduction in the price of transportation would also increase the number of miles traveled and energy consumed. But passenger miles traveled are not the same as vehicle miles traveled. The pooling of vehicles, encouraged by policy and the search for more revenue, follows from the proprietor's quest for efficiency via application of fleet logic. The massive inefficiency of today's transportation sector succumbs to these new business realities. The search for efficiency also finds its way to electrification.

Fleet logic will motivate increased utilization and higher fuel economy, with emissions savings as a spin-off benefit. It will also speed up depreciation of capital and turn proprietors into a powerful market force for innovation. Fleet operators will want cars that are reliable, economical to operate, and safe. Fast amortization means that compared to products owned by consumers, those managed and owned by proprietors and utilized by patrons will be newer, better maintained, and more diverse. Patrons will own fewer major assets, which will make them more mobile. Proprietors will deploy labor and capital more efficiently, and patrons will utilize them more efficiently. It is an efficiency revolution based not solely on technology but also on a business model.

A $5.4 Trillion Threat and Opportunity

This promise of increased utilization through pooling, fleet logic, and automation is not assured. Automakers are at the epicenter. They are key actors in how this plays out. They see two threats, both related to pooling.

First, if cars are used for pooling, fewer cars will be sold because people will be forsaking car ownership in favor of "chauffeured" ridehailing and eventually automated cars. More use of pooling, which would occur with automation, would further reduce sales. Second, if the customer is a mobility service company seeking low cost, it will be less tantalized by the high-value extras that individuals typically add while in the showroom; these add-ons are what generate the most profit for automakers.

Automakers could resist this pooling attraction by loading more technology and features onto cars and targeting them to affluent individuals. Premium cars earn premium profits. Automakers could specialize their technology offerings to strengthen their brand and target niche markets— such as enhancing the driving experience, offering plush comfort, or serving the cognitive limitations of the elderly. Automated technology is already becoming a key selling point. Over time, they could stay focused on pulling automation technology downstream to the mass market.

With this strategy, automakers would be sustaining their hundred-year-old business model of selling stuff. Most cars are bought by affluent individuals, so this premium technology strategy could succeed for decades. This strategy would also give automakers a better chance of maintaining control of the technology in their vehicles—and thereby profits.

But this focus on selling computers on wheels to individual buyers would expose automakers to the risk that software developers such as Apple, Google, Qualcomm, and Microsoft will capture most of the added value. When IBM allowed Microsoft to control the software in personal computers, Microsoft ended up earning profit margins greater than 30 percent, while IBM barely survived.

The competing narrative and business model is that instead of staying focused on individual buyers, automakers will morph into mobility service companies—proprietors that provide services to patrons rather than physical products. In this future, shared mobility with or without automation can be seen as not just a threat but also an opportunity. Indeed, Price Waterhouse Coopers predicts that between today and 2030, the share of automotive industry profits from new vehicle sales will fall from 41 percent to only 29 percent. But the overall size of that pie will grow from $400 billion to $600 billion. Mobility services, they say, will grow to account for around 20 percent of industry profits.[16]

In 2016, Ford began to reposition itself as a mobility company, based on the fact that the global market for new automobiles is about $2.3 trillion but the market for mobility is more than twice as big at $5.4 trillion.[17] Bill Ford, grandson of Henry Ford and executive chairman of the Ford Motor Company, said at the 2017 Automotive News World Congress, "I think this is a huge opportunity for companies. It's going to create revenue streams and business opportunities the likes of which we've never seen."[18] The company has established a new Smart Mobility business unit with the rather broad purview of "changing the way the world moves to make people's lives better" and with a focus on mobility, connectivity, autonomous vehicles, customer experience, and data analytics.[19] The business unit was established under CEO Mark Fields, who said, "It's not about moving from an old business to a new business. It's moving to a bigger business."[20] As so often happens, Fields was made obsolete by a revolution he himself championed. In May 2017, Ford replaced Fields with Jim Hackett, who had been chairman of the Smart Mobility unit.

Other automakers are also investing in new mobility schemes. Many are investing in multiple competing business models. GM is building a new mobility-focused brand it calls Maven. One initiative by Maven is to provide a "seamless" carsharing experience through an app and hourly car rentals.[21] In 2016, GM invested $500 million in Lyft with the intent of rolling out autonomous all-electric Chevy Bolts for use by Lyft. GM is also partnering with Lyft to provide short-term car rentals to its drivers. In addition, Maven has announced a similar pilot for short-term rentals with Lyft's competitor, Uber.[22] Although GM is hedging its bets in terms of its commitment to mobility services and who it partners with, its direction is clear. According to GM CEO Mary Barra, "We have the opportunity and the responsibility to create a new model of personal transportation."[23]

Daimler has invested in short-haul, one-way car rentals through its Car2Go subsidiary since 2008. The BMW Group followed with its Drive-Now in Europe and ReachNow in the United States. According to John Branding, BMW's director for government and external affairs in Washington, DC, the goal is to provide a BMW-grade experience throughout the entire mobility value chain: "After taking a rough cab ride to the airport the other day, I asked myself, 'Would I be willing to pay more for a BMW experience?' . . . Absolutely." He added, "For us, end-to-end premium mobility is the ultimate goal."[24]

The business models by which these companies will actually operate and profit are hazy at best. On revenues of $6.5 billion, Uber lost almost $3 billion in 2016[25] (though much of the revenue was used to finance expansion around the world). Clearly that is not sustainable. Another business model has been proposed by Tesla.[26] This scheme would see individuals purchase autonomy-enabled vehicles that would function as taxis when not in use by the owner. It is an interesting hybridization of the concept of consumer, producer, proprietor, and patron. But it's bound to be complicated. Simpler is the idea that proprietors will own the capital outright.

The New Symbiosis between Automakers and Ridehailing Companies

Automakers will increasingly find themselves in a complicated relationship with ridehailing companies like Lyft and Uber. The latter will be competitors but also customers. In the next few years, sales to drivers for Lyft, Uber, and others will almost certainly offset declines in car purchases by users of the services.[27] As shared mobility services expand, auto manufacturers will likely start producing customized vehicles for ridehailing companies—even as their own mobility services compete head-to-head.

Today, Lyft and Uber have almost no hard assets and are essentially Internet-platform companies. Managing and owning vehicles clashes with their current business model. But once level-5 driverless cars come on the scene, they will have little choice. It appears inevitable that they will embrace a new business model that involves ownership of vehicles. A huge part of the success of demand-responsive, app-based ridehailing companies will be their ability to profit as proprietors of capital, which will mean owning and operating a fleet of automated electric cars. So just as self-driving cars have pulled Detroit into the Silicon Valley game, the power of fleet ownership will likely force ridehailing companies into partnership with manufacturers.

As ridehailing companies build their customer base to support pooled services—which require a much larger base than glorified taxi services like UberX—they will use their purchasing power to push for vehicles that are specifically optimized for their needs. "We haven't started in on what the vehicle is going to look like. But it probably doesn't look like a Toyota Camry today," says Uber's Andrew Salzberg.[28]

The vehicles will be more passenger-centric—aimed at creating a comfortable passenger experience—as opposed to today's driver-centric cars. It's possible they will include movable partitions or compartments that will allow privacy to poolers who desire it, not unlike barriers between business-class seats on airlines. Since passengers will no longer need to concentrate on driving, there will be an increased focus on entertainment and connectivity within driverless vehicles. Immersive video games, video conferencing, or high-speed Internet will become commonplace for many platforms.

In 2016, recognizing this emerging integration of car making and mobility services, Daimler, along with US investor Plug and Play and Stuttgart University, founded an innovation hub in Stuttgart housing a number of auto-focused start-ups.[29] These start-ups, including one developing an in-car audio system that can wirelessly transmit sound to the ears of individual passengers without anyone else hearing the audio, aim to explore ideas related to future mobility. It is part of Daimler's goal to provide young firms access to the automakers' resources and knowledge while gleaning promising ideas that can be turned into products that meet the needs of the new mobility marketplace.

Lease deals are another way auto manufacturers might partner with mobility companies. In 2017, Paris-based BlaBlaCar, which focuses on providing seats for long trips, unveiled plans to offer its drivers cheap monthly leases for Opel cars, beginning in France. The start-up's nine million drivers buy an estimated 1.3 million new cars each year. BlaBlaCar CEO Nicolas Brusson declared, "We can pioneer a new approach to car ownership based on usage."[30]

All this points toward a messy era of transition where automakers, ridehailing companies, software companies, and other innovators will operate in cooperation and competition simultaneously. Over the decades, they will grope their way through a convoluted series of new partnerships and alliances.

Monetizing the Real Internet of Things

Regardless of whether they own or manage a fleet, proprietors are going to utilize and monetize their assets to the fullest. That means doing more than just filling vehicles—cars and vans—with the maximum number of passengers. It means finding new ways to unlock value based on the communications, processing capabilities, and physical location of those assets and the patrons who utilize them. The cumulative result will be yet another fundamental shift in the business model for existing manufacturers and tech companies and the emergence of a vast frontier of new possibilities that could not have existed in the consumer-producer transportation system of 2017.

Toyota, for one, is looking for ways to tap the Internet of Things for higher margins. In April 2016, the company created a new service that lets owners use vehicles like mobile phones. The leader of the new Connected Company business unit, Shigeki Tomoyama, said that "as vehicles become information devices, the car business's value added will be shifting to the cloud in the future."[31] As a first step, Toyota unveiled Smart Key Box, a gadget that simplifies carsharing by turning a person's mobile phone into a car key.

When driverless vehicles arrive, they will dramatically lower the cost of conveying physical things and create opportunities for integrating entertainment and business into the transportation experience. Driverless trucks will deliver bulk goods between businesses and warehouses. Smaller driverless trucks and cars will deliver parcels to staging depots and homes. Inside passenger vehicles, entertainment and retail will turn into major revenue streams. More lucrative will be the ability to monetize the process of physically bringing people together through dynamic pooling. Once the number of poolers becomes high enough, networking and dating services will flourish. All these services will also act as a platform for data gathering—helping companies build more sophisticated patron profiles and better understand tastes.

Will Automakers Make the Leap?

Mobility is poised to disrupt the business of car companies. This new wave of creative destruction is likely to devastate traditional manufacturing and energy markets. At the same time, mobility is the perfect hedge. While the profits of auto manufacturing might stagnate, the total revenues in the mobility space might well dwarf anything that currently exists. That's why Toyota, Ford, Saudi Aramco, and many others are already betting on mobility, with investments in the billions. The rising masses of mobility patrons will create an opportunity for proprietors at a scale the world has seldom seen. The repercussions for our society and labor sectors will be far reaching.

Shared mobility and automation innovations will compel producers to become either the proprietors or the victims of a massive sectoral culling. Proprietors will sell trips and services rather than horsepower, and those services will be delivered more efficiently than ever before. Patrons will consume these services in new configurations but with greater consciousness of the incremental cost—and none too soon. Mobile computing and human-dependent transportation are a toxic cocktail from a safety standpoint, leading to distracted driving and more car crashes. Car sales are booming around the world, together with associated greenhouse gas emissions. The current producer-consumer economic paradigm is increasingly untenable from an environmental perspective.

But pooling and automation are not guaranteed to emerge victorious. Many automakers might decide that they are better off stopping the march to automation at level-4 self-driving cars sold to individuals. Moving to fully driverless level-5 cars might be less profitable because hardware and software inside the vehicle sold by technology companies would become far more profitable than car making. The cars themselves might well become more commoditized because of the intense market pressure to reduce cost. That commoditized future is a specter that haunts automakers. And it's why they might try to tap the brakes on an industry shift toward level-5 driverless cars and pooling. There are many ways they could do so. Manufacturers might discourage regulatory approvals of level-5 cars, or they could create local roadblocks to their usage.

Still, there is much hope for the future. New technology, shifting consumer demands, and some supportive regulations already have us on the road toward a more technologically dynamic, less environmentally destructive, and safer transportation system. Getting there depends largely on policy leadership over the coming five to ten years—and a willingness of societies to bid farewell to stuff and fully embrace the power of proprietors and patrons.

CHAPTER 8

The Dark Horse: Will China Win the Electric, Automated, Shared Mobility Race?

Michael J. Dunne

Given its enormous pollution and congestion problems, along with the desire of its leaders to hold onto centralized power, China has tremendous motivation to lead the world in adopting electric, automated, shared vehicles. But it does not have a coordinated master plan and will likely pursue a zigzag path.

CHINESE LEADERS FIND THEMSELVES CONFRONTED WITH A massive, self-inflicted threat: the worst air quality in the world. Today's educated middle-class urban dwellers are angry about the pollution and the associated health risks. This constituency poses a real danger to those in power. As a result, Chinese president Xi Jinping and his comrades on the Politburo feel an urgent need to find solutions, including making full use of new technologies to combat pollution and congestion.

Will China be able to harness electric vehicles (EVs), automation, and shared mobility to sharply reduce its smothering air pollution and traffic congestion? Dramatic improvements are possible, but Chinese traditions in governance suggest that they will come not in precise steps or straight lines but along a zigzag path. China is not like Japan or Germany, where government officials develop highly detailed plans to reach specific goals within a tightly defined period of time. Chinese policy initiatives instead feature highly ambitious (often unrealistic) targets, vague timelines, tensions between national and local governments, and unclear accountability.

Further complicating the picture is the fact that China remains a partly reformed command economy. Powerful state enterprises still dominate large swathes of the economy, but smaller private companies move fast and are growing quickly. Leaders in Beijing feel challenged to strike a balance between supporting entrenched (largely noninnovative) state enterprises and highly creative private companies that are more difficult to control. How much Beijing officials are willing to back the private sector matters because private Chinese companies are clearly the ones driving innovation in EVs, automated vehicles (AVs), and shared mobility.

If it looks at times as if policies in China are contradictory or inconsistent, it is probably because they are. Think of Chinese policy making as a never-ending tug-of-war along several dimensions. It is true that China is a one-party system, but behind the homogeneous veneer is a tough and messy world of give and take. The central government spars with powerful city mayors over strategic direction. Powerful state enterprises vie with smaller, more agile and creative private enterprises for government funding and regulatory support. China carries on a tense hot-cold-hot-cold relationship with foreign companies. And even ministries within the central government joust for the upper hand. Officials in the Ministry of Industry and Information Technology have few friends in the Ministry of the Environment. No one aside from perhaps President Xi Jinping has the final word, and even he has to work very hard for it.

China's successful pursuit of the three revolutions depends on the extent to which leaders in Beijing can effect new policies in a smart and efficient manner to mobilize technology and consumer acceptance. The country will come to embrace the three revolutions, but the path will be neither straight nor tidy. Look for certain cities along the coast to take the lead in the three revolutions where and when it makes the most economic and political sense.

The Danger and the Opportunity

Two characters (危機, pronounced *weiji*) make up the Chinese word for "crisis." The first character, *wei*, means "danger." The second character, *ji*, means "opportunity." For China's leaders, the very real danger of pollution leading to social upheaval sits side by side with the very real opportunity for leadership in sustainable transportation.

Less than 2 percent of China's top five hundred cities meet the air pollution standards set by the World Health Organization (WHO).[1] Cities with the most extreme pollution line up west to east like a string of beads: Urumpqi, Lanzhou, Taiyuan, Xingtai, Shijiazhuang, Baoding, and Beijing. Cities like these issue air pollution red alerts for acute levels of smog and particulates. Schools close, factories are shuttered, and office workers stay home. Highways are closed and flights are postponed.

Many of China's urban centers are literally smothered in smog. According to the WHO, more people die from diseases related to air pollution in China than in any other country. Researchers at the respected Planck Institute in Germany estimate that contaminated air is responsible for 1.4 million premature deaths every year in China.[2]

Coal-burning plants and car emissions are the two main culprits, although in Beijing, dust blown in from the Gobi Desert during late winter and early spring windstorms is also a factor. Coal is the single largest contributor, responsible for 40 percent of the hazardous particulate matter found in Chinese skies.[3] Cars clog city streets and have come under particular scrutiny because of their sheer number. By 2020, China is expected to have more than 300 million vehicles on the road.

By comparison, the United States had 261 million vehicles in operation at the end of 2016.[4] The tremendous surge in vehicles threatens even worse pollution because Chinese vehicle emissions standards trail those of the United States and Europe by several years. And enforcement of standards is lax and inconsistent.

For centuries, those in power in China have had a deep-seated fear of *luan*, or social chaos. Given the serious and imminent political pressure related to air pollution, China has a strong incentive to quickly adopt and implement new sustainable transportation solutions. Chinese leaders understand that putting zero-emission vehicles (ZEVs) on the road is a powerful way of demonstrating to their ever-more-demanding middle-class citizens that the government is doing everything in its power to improve air quality. Federal and city agencies are devoting billions of dollars to subsidizing sales and building charging infrastructure. In 2010, the country set out strategic plans to be number one globally in EVs. Officials in Beijing declared specific production targets: five hundred thousand annually by 2015 and two million per year by 2020. And indeed, by the end of 2016, China had accomplished its goal of global leadership in EV production, turning out slightly more than five hundred thousand battery and plug-in hybrid cars, trucks, vans, and buses in that year alone—far ahead of any other nation.[5]

But whereas Chinese policy makers were forward-looking regarding EV production, they have been struggling to keep pace with the private sector when it comes to shared mobility and AV development. Didi Chuxing, the Uber of China, has achieved spectacular growth since its formation in 2012. It is now by far the world's largest on-demand ridehailing service. In 2017, the Chinese people were ordering more than twenty million Didi Chuxing rides every day. That works out to more than seven billion rides a year. Most of the growth came without regulatory guidelines. It was not until the summer of 2016, in fact, that Chinese regulators even managed to make the ridehailing business officially legal. Playing catch-up has given planners little time so far for strategic thinking about shared mobility systems.[6]

Chinese policy makers have been equally slow on rules for automated vehicles. It was not until the end of 2016 that the Ministry of Industry and Information Technology (MIIT), together with China's Society of Automotive Engineers (C-SAE), published a comprehensive and ambitious 450-page road map for technology leadership that set targets for phasing in automated vehicles over the next fifteen years.[7] The "Technology Road Map for Energy Saving and New Energy Vehicles" calls for driver-assist features to be in 50 percent of all new cars built in China by 2020. That number looks aggressive until you realize that by 2017, the majority of new cars in Europe, Japan, and the United States already came with driver-assist features (automatic braking, backup cameras, parking assist, lane departure warnings, and the like). The 2025 target is for 15 percent of new vehicles to be "highly automated" (not quite fully automated), and the 2030 target is for 10 percent of new vehicles to be "fully automated." Though the share might appear modest, it would translate into nearly four million vehicles.

The document also sets goals for greenhouse gas emissions and market share for so-called new-energy vehicles (NEVs, a category that includes EVs, plug-in hybrids, and hydrogen-powered cars) for the next fifteen years. The targets for emissions make clear that EVs will need to be a part of every automaker's portfolio in the future. NEVs should make up 7 percent of new vehicles in 2020, 20 percent in 2025, and 40 percent in 2030. China is expected to produce 33 million vehicles annually by 2020. Based on the targets, this would imply production of 2.3 million NEVs that year. By 2030, total vehicle production could go as high as 41 million,[8] implying production of 16 million NEVs.

Hit those numbers, and China will surely lead the world in vehicle electrification, along with shared mobility. But what about automation, and what about integrating them? Imagine for a moment that Chinese president Xi Jinping makes mastering the three revolutions a top national priority. The country pulls together to make Chinese cities and rural areas home to the most advanced, sustainable transportation system on the planet by 2030. Real breakthroughs in air quality, traffic congestion, and quality of life, unthinkable just a few years ago, are now suddenly within reach.

Several attributes for success are already in place. The country leads the world in EV production and ridehailing services. Hong Kong and Singapore, two "overseas Chinese cities," have already built some of the most advanced urban transport systems in the world. Hong Kong was ranked the world's most advanced city when it comes to efficient transportation, ahead of ninety-four other cities, by Mobility Index 2.0, a study by Arthur D. Little published in 2014.[9] Singapore was rated sixth in the same study. And some cities like Shanghai have managed to limit traffic congestion through vehicle purchase quotas and auctions for licenses. Other cities are populating their entire taxi fleets with EVs.

China's single-party power structure gives us more reason to believe that the country will lead the world in adopting shared mobility systems, powered by EVs, ridehailing apps, and automation. China at times appears to have the ability to wave a wand, mandate sweeping new directives, and reach sky-high goals that other nations can only dream of. Consider the astonishing speed and magnitude of the massive infrastructure projects completed between 2000 and 2014 as shown in table 8.1. The increases in capacity are jaw-dropping. Never before in the history of humankind has so much been built in such a short span of time. It's tempting to think that China will make a similar great leap in sustainable transportation.

Alas, if only China's mission of transformation via the modern triumvirate of electrics, shared mobility, and AVs were so simple. There are three significant obstacles to overcome: technical, market, and political.

Table 8.1. Massive infrastructure growth in China, 2000–2014

	2000	2014
Railways	60,000 km	120,000 km
Roadways	1.4 million km	4.5 million km
Highways	16,000 km	120,000 km
Solar panel capacity	19 MW	43,000 MW
Subways	400 km	3,000 km
Air traffic passengers	67 million	390 million
Vehicle production	625,000	22.1 million

Source: National Bureau of Statistics of China, 2000–2015; EU SME Centre, "The Automotive Market in China (2015 Update)," 4 June 2015.

The Technology Challenge

Let's start with the technology challenge. Although China is now the world's number one producer of EVs, it is not a leader in EV technology. The world's best hybrid and EV know-how still resides in the advanced markets of Japan, the United States, and Europe. And China's homegrown AV innovation capabilities are weak; it lacks the electronic, sensor technology, and software engineering expertise that has made Silicon Valley the epicenter of AV technology.

How to catch up? For decades, China's state enterprises have made futile attempts to secure technology from their global automotive joint venture partners. To make real headway in technology and quality, the Chinese government has had to loosen the terms of foreign participation and open the door to Internet giants.

Loosening the Terms of Foreign Participation

China has been quite insistent that Chinese companies play a leading role in the EV industry. In early 2017, the government showed some signs of opening up to greater foreign participation, but the ultimate aim seems to be for Chinese firms to keep control of the EV industry.

For China's EV industry to prosper, batteries must be inexpensive, safe, and reliably long lived. But local battery makers were not producing world-class batteries. Market leader BYD does produce affordable batteries in high volume, and some local battery makers are approaching world-class quality, but many are still falling short. Foreign investment and technology could upgrade battery quality.

Foreign ownership in battery production companies was capped at 50 percent until the summer of 2016, when Chinese regulators announced that foreign companies could now own up to 100 percent of local companies, provided they set up operations in one of three special trade zones. This is an important step in attracting vital investment and battery technology. Lifting the investment cap allows leading global battery makers such as LG Chem (South Korea), Panasonic (Japan), and Samsung (South Korea) to control manufacturing operations (and protect their intellectual property).[10] Will the investments really happen? The Chinese government maintains a final list of qualified suppliers, a so-called white list. Contrary to expectations, some Korean battery manufacturers did not secure a position on the white list in 2016, and as a result, they were not able to capture important contracts.

Similarly, between 2010 and 2016, China firmly insisted that any global automaker that wanted to sell EVs in China would have to build cars in partnership with a Chinese enterprise that owned 50 percent or more of the business and that the product would go to market under a Chinese brand name. Global automakers that elected to export directly to China were hit with stiff import duties and fees that rendered the vehicle price higher by 50 percent. A Tesla Model S selling for $80,000 in the United States, for example, would be showroom priced at $125,000 after import duties and fees. As a result, Tesla's sales in China have been limited. China's premium car market reached two million units in 2016, yet Tesla sales managed to climb to just ten thousand, or less than 1 percent. Still, this was enough to put the Tesla Model S among China's top twenty best-selling EVs in 2016, the only foreign model to be ranked there.[11] Foreign brands, as a whole, accounted for less than 3 percent of China's EV market, although they still account for the majority of combustion engine cars sold in China.

Central planners in Beijing had believed that foreign automakers and suppliers would be willing to share EV and battery technology in exchange for market access. Who could resist China, the world's largest market? That bet proved to be misguided. The aggressive technical-sharing rules over the previous five years had made investing in China's EV industry too risky and thus unpalatable for foreign automakers, who were content to continue building and selling conventional combustion engine cars and trucks inside China. German, Korean, Japanese, and American carmakers paid lip service to China's desire to lead the world in electrics, but they stayed largely on the sidelines, keeping a tight hold on their precious EV technology.

China finally understood that to get more advanced technology, it would need increased foreign participation. One striking illustration of how Chinese acquisition of American technology can accelerate China's EV industry is the Shanghai-GM joint venture's signing in January 2017 of a $1 billion contract with A123 Systems to supply batteries for future joint venture products. A123 Systems, an American MIT spinoff that had gone bankrupt, was acquired by China's Wanxiang Group in 2012 and produces inside China as well as in the United States.[12] China could develop batteries and EVs in relative isolation from the rest of the world, but the quality would be questionable. China continues to wrestle with ways to secure foreign technology while keeping overall control of the industry.

In September 2017, the Chinese government announced a new regulation that 3 percent of sales in China by global automakers must be "new energy vehicles" beginning in 2019, with the percentage ramping up after that.[13] VW, Mercedes, Audi, Tesla, and GM, among others, were reportedly engaged in talks to build EVs in China during the run-up to 2020.

Officials in Beijing came to understand that foreigners must be involved for China to develop and sustain global leadership in electrics—at least at the outset.

Opening the Door to Internet Giants

Arguably the single most important policy initiative for the three revolutions has been the decision to open the automotive industry to nontraditional automotive companies. For the first time ever, the Chinese government has invited private Internet companies, including LeEco, Alibaba, Tencent, and Baidu, to compete in China's automotive industry.

An unexpected benefit of inviting these players into the arena has been an infusion of new thinking about technology transfer. While state enterprises had been relying for a generation on their global automaker partners for transfer of technical know-how, firms like LeEco and Baidu ignored the tech-transfer option and quickly established wholly owned operations in California to access state-of-the-art talent and technology. And inviting new competition put real pressure on the incumbent (and sometimes lethargic) automotive state enterprises like Beijing Automotive, First Auto Works, and Dongfeng Motors. This new competition has inspired the state enterprises to establish venture capital and R&D bases in California too.

Perhaps the company with the best potential for autonomous vehicle leadership in China is search giant Baidu. Founded in 2004, Baidu employs more than forty-six thousand people, is listed on NASDAQ, and has annual revenues of more than $10 billion.[14] Baidu sees California as the quickest route to mastery over AV technology and has established a formidable research and development operation in Sunnyvale, California, just down the road from Apple. That center houses three distinct but connected research centers: the Big Data Lab, the Institute of Deep Learning, and the Silicon Valley Artificial Intelligence Lab. Baidu has established an important relationship with Nvidia, a leading

US computer graphics and deep learning company based in San Jose, California, to work on a new computing platform for AVs using artificial intelligence. Baidu also secured approval from the State of California in 2016 to conduct AV testing.

Still, its autonomous technology remains a few laps behind that of global leaders such as Google, Tesla, Israel's Mobileye, and Delphi, the former GM subsidiary. In a surprise move, Baidu announced in April 2017 that it was making its autonomous vehicle technology freely available to anyone who wants it under an initiative called Project Apollo.[15] It seemed to be hoping that by open sourcing its basic platform, it would attract contributions of new features from the community and the platform would be adopted more broadly and quickly. Indeed, Beijing Automotive featured a new model equipped with Baidu's autonomous drive capabilities at the April 2017 Shanghai Auto Show.[16]

Didi Chuxing is positioning itself as a serious rival, reportedly raising nearly $6 billion to invest in self-driving cars and opening a self-driving car lab in California in March 2017. By the same time, three other Chinese-funded companies with operations in California had announced significant advancements in the design and development of premium EVs and were investing in AV capabilities in California too.

- NextEV—Founded in 2014 by Chinese Internet billionaire William Li and based in San Jose, California, NextEV aims to remake transportation by building cars that are electric, autonomous, and connected.[17] In early 2017, the company unveiled the world's fastest electric car, the NIO EP9. NextEV separately invested $465 million in new facilities to produce premium electric cars in the eastern Chinese province of Jiangsu.[18]
- Faraday Future (FF)—FF is based in Gardena, California, and employs roughly a thousand people, many of whom came from Tesla, BMW, and other premier car companies. With funding from

LeEco's founder, Jia Yueting, FF revealed an impressive high-speed electric crossover called the FF91, complete with fully automated self-parking features, in early 2017. The company had plans to build a plant in North Las Vegas and begin production in 2018, but financial distress caused FF to put the project on hold in mid-2017. The vehicle was expected to be priced above $150,000. FF's parent, LeEco, had indicated plans to invest $3 billion in a greenfield facility in Huzhou City, Zhejiang province.[19]

- Lucid Motors—Founded in 2007 by ex-Tesla executives and with funding from China, Lucid is based in Menlo Park, California, and also has American and Japanese backers. In late 2016, Lucid introduced the Lucid Air, a luxury sedan designed to take on Mercedes, BMW, and Audi. Like the FF91, the Lucid Air is extremely quick, going from zero to sixty miles per hour in 2.5 seconds. The company has identified a future production site in Casa Grande, Arizona.[20]

Not all these firms will prevail in the intensely competitive car business, but the trail they have blazed into California opens huge new opportunities for China to catch up to its global rivals in a short time. Operating in California has been a strategic game changer. The speed with which Chinese companies in California have attracted capital and developed new vehicles has been breathtaking. Quite clearly, Chinese companies—just like those in Detroit, Europe, and Japan—are investing in California technology and talent to develop the full spectrum of electric and automated vehicle technology.

Once the cars are refined in the United States, they will most likely be built and sold in China too. China is now home to the world's largest luxury vehicle market, so it is not too hard to imagine Chinese-funded premium electric cars designed and developed in California becoming attractive to wealthy people in Shanghai, Beijing, and Shenzhen. And looking longer term, these highly advanced vehicles can be seen as part of a learning phase that will result in more and more high-quality, affordable EVs. Already Baidu has unveiled a new all-electric self-driving car for testing in China—a modified Chery EQ with the company's driverless technology installed. Baidu is looking to launch a public shuttle service in China using such vehicles by 2018.[21]

Chinese leaders, for their part, understand that developing AV technology at home would be a long and arduous road. It is far better to secure the technology directly from world-class technology centers in California. Once the technology comes to China, the government will be well placed to direct leading companies like Baidu on where and how to apply the technical advances.

Reluctant Consumers

Another impediment to quick progress toward a sustainable transport system built on the three revolutions is Chinese consumers themselves. Most are still highly reluctant to acquire EVs, citing range anxiety, uncertain resale values, a shortage of charging locations, and doubts about safety and reliability.[22] Further complicating the buying arena is the fact that Chinese carmakers do not have a reputation for building quality cars, let alone quality EVs.

The majority of the five hundred thousand NEVs sold in China in 2016 went to city-owned taxi fleets or transportation companies, or they were acquired in Beijing and Shanghai, cities with quotas and huge subsidies. In other words, the state helped the state reach state targets—never a good formula for sustained success.

To generate sales momentum, national and city governments have offered generous subsidies for NEV purchases, as high as $8,475 per vehicle.[23] Some cities, including Beijing and Shanghai, have also waived quotas and special licensing requirements in the hopes of driving higher EV sales. The concentrated cash subsidies and other favorable policies have paid off in terms of total sales, but the record deliveries have come with substantial costs.

For one thing, estimates are that subsidies amounted to $24,000 per plug-in hybrid vehicle and $29,000 per battery EV in 2016 in some regions, far more than in the United States.[24] On top of that, reports of fraud are widespread. Several Chinese automakers are under investigation for taking subsidies for sales of EVs to "fake customers" in 2015.[25] A company would sell a bus or car to a subsidiary company on paper, collect the cash subsidy, and then have the subsidiary company later "return" the car. It seems that actors at the bottom were finding ways to circumvent the top's policies.

How will Chinese planners address the fact that they find themselves saddled with a high-volume, meager-technology EV industry of uncertain quality that has been plagued with fraud? The government's actions to open up to foreigners and invite nontraditional automotive firms to enter the arena are promising signs for the future. Customer reluctance can be overcome as the quality and convenience of EVs improve.

The 2016 "Technology Road Map for Energy Saving and New Energy Vehicles," mentioned earlier, promises some steps to increase consumer demand. The blueprint proposes that electric cars will target the compact car market, while plug-in hybrids will pull in people who are looking for larger cars. The document also sets targets for increasing battery range and building charging facilities across the country. The goal is to have at least five million charging posts in the country by 2020, no fewer than twenty million by 2025, and no fewer than eighty million by 2030, to match the number of new-energy cars in the country. Another goal is to make electric cars affordable without government subsidies.[26]

But even if EVs aren't yet going over big, the sensational growth of China's ridehailing industry seems to indicate that Chinese consumers are open to the concept of mobility as a service. Uber entered the China market in 2014 with great expectations, but after losing more than $1 billion in a war of attrition with Didi Chuxing, Uber agreed to be absorbed by Didi in the summer of 2016.[27] This was about the same time that China officially recognized ridehailing apps as legal businesses. Bloomberg estimated that by 2017, Didi had 88 percent of all the ridehailing users in the world (about four hundred million) and 81 percent of all the drivers (fourteen million) and had raised 38 percent of all the financing ($12 billion). In contrast, Uber, the second largest ridehailing company in the world, had 9 percent of all the riders and drivers and had raised 37 percent of all funding, while Lyft, the third largest ridehailing company, was at 1 percent of worldwide users and drivers and 8 percent of total funding (though Lyft operates only in the United States).[28]

Clearly, the Chinese are embracing shared mobility more quickly and on a more dramatic scale than Americans and Europeans. The sharp takeoff in ridehailing popularity is attributable in part to China's suffocating congestion and related inconveniences. Cars have been important status symbols for the newly affluent and for educated city professionals, but the joy of being seen in a new Ford SUV or Honda sedan is increasingly offset by the frustrations of traffic jams and hard-to-find parking. This trend is most visible in dense cities like Shanghai, Shenzhen, and Dalian. In Beijing, the conversation among friends often revolves around the best way to get to destinations without having to drive. Increasingly, as Didi's record numbers indicate, the Chinese are opting for ridehailing services.

But let's remember that ridehailing is not the same as ridepooling (where you share a ride with a stranger). To be a truly effective transportation solution, ridehailing services should feature a high degree of pooling. How comfortable the Chinese will get with pooling remains to be seen. Certainly, it is already common for Chinese friends and office colleagues to share a Didi vehicle, which amounts to carpooling (where you share a ride with someone you know). Chinese urban professionals might be prepared to leave their cars behind in exchange for less driving and parking hassle, but China does not have a ridepooling culture—yet. Still, during the 2016 Chinese New Year holiday, more than one million commuters chose pooling as a way to get home when train, bus, and air tickets were sold out.[29]

Bikesharing is also part of China's urban mobility. In 1980, an estimated 63 percent of Chinese commuters got to work on a bike. I traveled by bike to my first job in downtown Beijing in 1988. Taxis were scarce, and private cars were almost nonexistent. Today, cycling's share of total commuters has dropped to just 12 percent. Chinese start-ups are hoping to reverse the trend. Ofo, one of China's leading bikesharing companies, was valued at $500 million in early 2017, with 250,000 brightly colored bikes on Chinese streets. Its competitor, Mobike, had placed more than 400,000 bikes in the major cities of Beijing, Shanghai, Guangzhou, and Shenzhen.[30] While this number looks promising, it's still early in the growth curve. And an obvious stumbling block to a bikesharing revolution is that those who are pedaling suffer direct exposure to dangerous particulate matter caused by coal-burning plants, dust, and vehicle exhaust, not to mention facing safety risks in traffic. Shanghai does a very good job of providing bike lanes, and Beijing and Guangzhou also offer some bike lanes.

Central and Local Powers Jockeying for Leverage

A third and probably the most important impediment to continuing surges in shared mobility is China's complex political economic system. Chinese regulations and plans that involve multiple parties at the central and local levels tend to take a circuitous route. Local and central powers constantly jostle for sway, leverage, and the final word. There is a classic Chinese adage that captures the essence of the struggle: "The top makes the policy, the bottom takes countermeasures" (上有政策下有對策。; *Shang you zheng ce, xia you dui ce*). For many directives emanating from the central government planning ministries in Beijing, a corresponding counteraction comes from leaders at the province or city level as each party works to protect turf. It's messy. It's muddled. And it's how China works.

As a result, China's embrace of electrics, ridesharing, and AVs will be highly uneven across cities and sections of the country. Look for the wealthier coastal cities like Shenzhen and Shanghai to take the lead, while second- and third-tier cities join the transformation anytime from five to fifteen years later.

During Mao Zedong's rule (1949–76), most power was concentrated in Beijing. But since Deng Xiaoping ushered in China's post-1970s era of reform and opening, cities have gained tremendous degrees of power and autonomy. Cities are essentially responsible for their own economic success. When central government policies are not aligned with local economic interests, Chinese mayors can stonewall and obfuscate with the best of them. The national government might want to promote more efficient transport systems, but powerful cities and provinces have other priorities: investment, employment, and tax revenues. It is nearly impossible to have a comprehensive master plan when the constituent cities have their own ideas about what is best.

Let's take the highly progressive city of Shanghai as an example. Think of Shanghai as Shanghai Inc., with the mayor acting as chairman and CEO. Shanghai state enterprise owns both the nation's largest automotive manufacturing operations (Shanghai Automotive Industry Corporation [SAIC]) and the city's taxi fleet. For an integrated sustainable transportation system to gain traction, the central government must first convince the mayor of Shanghai that no jobs will be lost. As important as it might be for the central government to solve air-quality problems, it is far more important for most city leaders to sustain and add jobs.

SAIC employs hundreds of thousands of people, generating profits and important tax revenues for the city. Shanghai's more than sixty thousand taxis create employment for several hundred thousand residents.[31]

The Shanghai Model for Sustainable Transport

Think of Shanghai as the New York City of China. It has an enormous popu-lation concentrated within a limited geographic area. The bustling metropolis of more than thirty million residents is home to the nation's tallest skyscrapers. It also has a long-standing reputation as a global business and financial center. Shanghai is beset by the same terrible air pollution that plagues China's other major cities, but its leaders have worked harder than those in most other cities to solve its traffic congestion and thus limit transportation's contribution to air pollution.

Shanghai invests heavily in new transportation infrastructure while also tamping down on excessive traffic through targeted regulations. The city addresses most of its present transport demands with a massive subway sys-tem. Between 2000 and 2015, Shanghai spent billions of dollars building more than 300 miles of subway lines. By 2020, that total will climb to a projected 530 miles, which will make Shanghai home to the world's largest urban rail network. An estimated ten million people commute to work using the Shang-hai metro every day.[32]

Since the early 2000s, Shanghai has placed a strict quota on the number of cars that can be purchased by private buyers. The number fluctuates around ten thousand cars per month. To buy a car, residents must also purchase a license through a blind auction. The auction price ranges between $12,000 and $15,000, depending on the number of applicants in a given month. The high price for a license serves to diminish the appetite of some would-be buyers and provides extra revenue the city puts toward public infrastructure.

And these measures are paying off, at least in terms of reducing congestion. In a 2016 study, Shanghai was not listed in the top ten most congested cities in the country (Beijing was third).[33] While Beijing and Shanghai have roughly the same number of residents, Shanghai has half as many vehicles in operation. Its streets are full of cars, but the traffic moves along. In Beijing, on the other hand, getting from downtown to the airport, seventeen miles northeast of the city center, can take more than two hours.

City planners are also working hard to drive demand toward EVs: buyers of electric cars are exempt from the city of Shanghai's special license fee. EV demand in Shanghai has not yet taken off due to residents' preference for large, high-quality vehicles, but that might change as Audi, Mercedes, and BMW prepare to produce more EVs inside China in the coming years. Meanwhile, bikesharing is big in Shanghai, where bikes with embedded GPS chips can be left anywhere in the city and located with a bike-hailing app.

Shanghai's own SAIC builds millions of cars a year (99 percent of which stay in China) in partnership with GM and Volkswagen. SAIC formed a $160 million partnership with Alibaba in 2016 to develop connected cars, understood to be the first step in a broader alliance to develop AVs. We can expect the City of Shanghai to introduce these cars to its transport network once the technology is proven safe and reliable.

This means that when Didi Chuxing acquired a license to operate in Shanghai starting in 2015, it found itself in the unenviable position of fighting the mayor's office. As ridehailing in Shanghai enjoys increasing popularity because it offers residents a way to avoid the hard work of finding a parking space for their own cars, it has taken a significant chunk of revenues away from the taxi fleets. In late 2016, the City of Shanghai announced new compliance and licensing rules that could dampen growth of the ridehailing business, at least in the short term.

If Shanghai veers toward protectionism and slow-walks innovation, what can be expected from other less-developed Chinese cities? Shanghai will grow more progressive, but it will be on the mayor's own terms and timing.

On the other hand, some municipalities are eager to sprint ahead, even if not part of a comprehensive national plan. Wuhu, a city of three million in central Anhui province, aims to be the first Chinese city to have only AVs run on its roads. Baidu is now joining hands with Chery, a state-owned automaker based in Wuhu, to build and test automated electric cars. They aspire to make the city's entire population of vehicles automated and electric by 2022.[34] "We are trying to give the experience and data to the central government so they can see the benefit, and that will make it easier for us to push to other cities in China," said Wang Jing, head of autonomous cars at Baidu.[35]

And then you have the central government pushing back. In 2016, authorities in Beijing placed a temporary ban on AV testing on Chinese roads. This action was intended to ensure safety on Chinese highways but has had the indirect impact of stalling AV progress within China. Some cities might soon be designated as test centers for AVs. This approach is consistent with China's tendency to test new things on a limited scale before rolling them out nationwide.

The bottom line is that we can expect China's embrace of the three revolutions to be marked by a fair share of stops and starts, inconsistencies, overhauls, reversals, contradictions, breakthroughs, and setbacks. Zigzagging its way to goals with little refinement or evidence of unity is China's most likely path forward. On the other hand, central government planners might once again have more clout if President Xi Jinping is able to recentralize power in Beijing, which he is working hard to do. Concentrated power could accelerate adoption of electrics, AVs, and shared mobility.

China in 2030: Projecting the Three Revolutions

Even if the road ahead is filled with twists and turns, it is likely that China will be producing electric, connected, and automated vehicles at world-class levels of quality and sophistication by 2030. This is particularly true with China's newfound direct access to technologies in California and its seeming willingness to open up to more foreign participation. China is now the world's leading importer of oil and thus has ever-increasing incentive to move toward electrics quickly.

And now that China has officially sanctioned ridehailing companies as legitimate businesses, planners can start to look at ways shared mobility might improve traffic patterns and reduce congestion, or the total number of cars on the road. The most likely direction is for cities to develop and possibly mandate shared AV routes, starting with buses, commercial vans, and fixed-route taxis. Chinese city officials are already grappling with how to benefit from ridehailing, given that some cities (including Shanghai, Beijing, and Guangzhou) own the taxi fleets that are under pressure from ridehailing apps. Endorsing companies like Didi means revenue losses in the short term. Look for the central government to strongly encourage cities to find ways to make shared mobility work.

Another reason to be optimistic about China's growth in ridehailing is the central government's apparent support for Didi as a monopoly or near-monopoly operation. When Didi absorbed Uber in the summer of 2016, Didi suddenly controlled 80 percent of the Chinese ridehailing market. An antitrust motion was filed, but the Chinese government quickly dismissed it. This indicates that the central planners are comfortable with Didi playing a dominant role in the industry. They anticipate having leverage to steer Didi in certain directions at certain times, much as they do with Baidu (the Google of China) and Weibo (the Twitter of China), which enjoy similar quasi-monopoly positions.

Real breakthroughs in China tend to happen at the city level, not nationally. And Chinese cities can be classified into at least five different tiers of economic development and incomes. It follows that the three revolutions will build across China in waves. By 2030, it is definitely possible to imagine the concentrated coastal cities of Shanghai, Xiamen, Shantou, Dalian, Fuzhou, Hangzhou, Nanjing, and Qingdao being the pioneers in adopting the three revolutions. Focusing efforts on the high-profile cities first will also serve to alleviate urban citizen pressure to improve air quality. Other cities will follow in increments to 2040 and 2050.

As with many policy initiatives in China, the vision starts with the central government. But it is the cities that almost always drive successful implementation. China's three revolutions will come in fits and starts in select cities along the coast. Come to think of it, you could say the very same thing about America and not be too far off.

Pooling Is the Answer

THE ANSWER IS POOLING. IF THE QUESTION is how to ameliorate traffic congestion, the answer is pooling. If it's how to reduce climate change, still pooling. Social equity? Also pooling. Soaring transportation infrastructure costs? Pooling! What to do about the potential negative effects of automated vehicles (AVs)? Pooling. Going forward, pooling must be the principal focus of our thinking and actions related to transportation.

Vehicle electrification brings important energy and environmental benefits, but the key to sustainable cities and transportation clearly is pooling. The energy and environmental benefits are multiplied, and many new benefits accrue—including less land required for parking, greater access by carless and mobility-disadvantaged travelers, reduced need for roads, and less cost to individuals and to society. As the three revolutions begin to converge, pooling of rides is key to the dream scenario we painted at the start of this book.

When we say pooling, we mean filling the empty seats in all our vehicles—cars, vans, buses, and rail (with bicycles exempted for obvious reasons). We mean encouraging greater use of Lyft Line, UberPool, microtransit, and conventional transit. For conventional transit, increased utilization means repositioning buses to serve dense travel routes, where they perform best. For ridehailing companies, it means shifting the focus to multiple riders and away from single-passenger taxi-like services. For microtransit, it means enhancing and shifting transportation funding to support demand-responsive, app-based, van-like services.

The good news is that ridehailing companies such as Lyft and Uber are already convinced. This is not a matter of pushing a boulder uphill; they are on board. They are certain that riders are so price sensitive that if fares are lowered even a little, many more will ride. If more people use their service, they earn more revenue and thus more profit. In every discussion I've had with Lyft and Uber executives, not a single one has resisted this argument. In fact, they enthusiastically endorse it. They proclaim this to be their goal. Here is a case where business models and the public interest are aligned. Pooling is unequivocally good from a societal perspective.

The second key to the dream scenario, beyond pooling, is vehicle electrification. If vehicles used for pooling are electrified, the energy and climate benefits are multiplied. And as electricity is decarbonized—made from renewable energy—the use of electric vehicles (EVs) will greatly reduce petroleum use, local air pollution, and greenhouse gas emissions. The same will be true for fuel cell EVs operating on increasingly low-carbon hydrogen.

And third, if these pooled EVs are automated, the benefits are even further enhanced. The cost per vehicle mile drops dramatically, from $0.57 for a conventional individually owned new car to $0.20 per mile or less (as shown earlier in figure 1.2), even assuming the automated vehicle costs $10,000 more than a conventional gasoline-powered car. It is far less expensive because the automated car is operated far more intensively—twelve hours or more instead of the typical one hour per day. The calculation of $0.20 assumes one hundred thousand miles per year, rather than today's average for individuals of fifteen thousand miles. If this automated EV has two riders, the cost per passenger is cut in half to $0.10 per mile.

Battery electric and hydrogen fuel cell EVs are a good match with automation because they have lower maintenance and energy costs than combustion engine vehicles. When the vehicles are used intensively, these lower operating costs more than offset the higher purchase cost. When vehicles are operated ten thousand to fifteen thousand miles per year, as they are now, the additional purchase cost of the vehicle plays a big role in the total cost of owning and operating the vehicle. When vehicles are driven forty thousand or even one hundred thousand miles per year, fuel costs and maintenance become far more critical. EVs, with far fewer moving parts than combustion engine vehicles, are also more reliable—another advantage that becomes more valuable as the vehicles are used more intensively. Reliability of electrified vehicles is still further enhanced when combined with automation. Because automated cars rely on electrical devices, the car's entire architecture is built around electric motors and devices—which are more reliable than mechanical and hydraulic devices.

If the automated EVs carry multiple riders, the debate over the desirability of AVs simply evaporates. Automated, shared EVs will have nothing but positive consequences for energy use, emissions, parking, and traffic congestion.

How Do We Get There?

We are just beginning. Lyft and Uber began rolling out pooling services in 2014, but these were only in thirty large metropolitan areas globally by the end of 2017. Microtransit was struggling, with no large pilot projects successful and some having already failed. Travel choices were still very limited. How do we get from here to there? The challenge is twofold: behavior and policy. Travelers need to embrace pooling, and policy needs to encourage it. Pooling depends on scale. A large volume of riders is needed for pooling services to become efficient.

Human behavior is central. We'll need experiments to determine which people under which conditions will embrace pooling services, and then we'll need to enact business strategies and policies to create and support those conditions. Will riders insist on control over who they share rides with? Will they want barriers between seats, as airlines have done with business-class seats? What type of privacy rules might be implemented that are legal and effective? Pooling will not appeal to all people all the time. But clearly many of us, perhaps most, would be willing to share rides if doing so were cheap, safe, easy, and reliable. After all, wouldn't most of us prefer to be chauffeured? For many of us, lower fares are the key. For others, security. Others require high reliability. And still others insist on near-instant availability.

Some will never be won over, at least for particular trips and at particular stages of their lives. For moms and dads with multiple kids, giving up their minivan would be painful. For sports enthusiasts who regularly surf, golf, ski, or kayak, having a vehicle to transport their equipment is key. But for most trips, most people could share rides in vehicles operated by mobility service companies—and would be amenable to doing so if the price and other factors were right.

A second big consideration, beyond the willingness of people to share rides and relinquish car ownership, is automaker intent. How enthusiastic and how rapid will automakers be in transitioning to passenger-centric cars—cars controlled by a robot and designed for passenger comfort? Automakers are conservative for good reason. One hiccup can lead to lawsuits and payouts that amount to billions of dollars. A faulty radar device, airbag, or wire that results in deaths is enough to bankrupt even a medium-sized car company. Failures of new technology elicit media attention. Automated cars that crash or even suffer fender benders rate headlines. A sour experience turns off customers and stockholders, many of whom are already skeptical of automation and electrification and new business models.

In a strategic sense, the path forward is as follows: first, create choice. Choice is a necessary precondition. If choice doesn't exist, travelers—car owners—see efforts to restrict car use as punitive. And indeed, they are correct. The vast majority of people in the United States, and increasingly elsewhere, depend fully on cars. They don't have a good alternative to the vehicle in their driveway. But with choice, they *will* be more receptive—as both voters and travelers—to actions that encourage pooling and discourage driving, resulting in a future with fewer vehicle miles traveled but *more* passenger miles traveled. It is a future of greater mobility. Travel will be cheaper and, in many ways, easier.

Choice starts with the glorified taxis of Uber, Lyft, and others, as well as carsharing and bikesharing, but politicians and others must commit immediately to a more expansive vision, a future framed by pooling. They need to signal to ridehailing companies, transit operators, car manufacturers, advocacy groups, and travelers that pooling (and electrification) is the future. Travelers must believe that single-occupant vehicle trips in combustion engine vehicles are headed toward the dustbin of history—or at least toward a sharply diminished role.

Policy makers must overcome concerns about current taxi-like ridehailing services. Yes, they cause small increases in overall vehicle use, and yes, they divert some trips away from cash-strapped transit—in effect providing large private benefits to individuals but few or no societal benefits. And yes, their Silicon Valley culture, especially Uber's, can sometimes be offensive. But politicians and government leaders must recognize that these companies are an essential stepping-stone to a better transportation future.

At the same time, Uber and Lyft must redouble their efforts to expand their Lyft Line and UberPool services. And transit operators must overcome their conservative bent and partner with ridehailing companies and microtransit start-ups. Local and state governments will need to encourage and help public transit operators as they gingerly partner with private mobility companies. As the image and profile of pooling increases—and as travelers become more comfortable with the notion that they have choice—local, regional, and state leaders can gradually enact a continuing stream of incentives for pooling and eventually introduce disincentives for single-occupant driving.

The next step is for regulators, using data and science, to determine when driverless vehicles are safe. Maintain incentives for pooling and add automation to the picture, and the path to the three revolutions dream scenario is clear. Nothing more will need to be done except to accelerate all initiatives related to reducing parking spaces and channeling electrified AVs into dedicated lanes, resulting in less congestion and higher speeds.

No-Regrets Policy Strategies

Potholes and detours are everywhere on this path to the future. How fast and in what ways technologies will evolve is hard to forecast. So is traveler response to new technologies and services. One obvious strategy is to greatly expand research on and demonstrations of the most promising innovations, together with studies of how travelers might respond.

As for policy, when contemplating major changes, one should emphasize market- and performance-based approaches. Prescriptive policies should be avoided because the future is too uncertain and unintended consequences are too likely. No one is omniscient, not even the smartest researchers who have dedicated their careers to transportation, energy, and the environment—and certainly not government policy makers. Much humility is called for. Thus policies are needed that are robust and flexible, able to accommodate unforeseen twists and turns, and durable so that industry and consumers can make investments and decisions with some amount of certainty about the future—or at least, the government's role in that future.

Put another way, which incentives and disincentives are most effective at motivating both pooling and electrification? Which are most effective at guiding companies and consumers to a future of shared, electric AVs?

We have suggested many specific policy ideas throughout this book. All are premised on accelerating the transition to pooled EVs and eventually transitioning to pooled automated EVs. These policies involve creating incentives and rules that encourage mobility service providers to promote pooling and use EVs and that inspire travelers to use those pooling services. They include market- and performance-based policies and regulations to steer automakers toward designing their vehicles for pooling and powering their vehicles with electricity and hydrogen. They include policies to encourage public transit operators and mobility service companies to collaborate in providing more access and service at lower cost. And they include policies to urge local governments to redesign cities with less parking and more space for walking and biking.

Final Thoughts

This book presents a hopeful vision of our transportation future and suggests how we might steer toward it. While the challenges in achieving our dream scenario are daunting, pooling is a good starting point. It provides the organizing framework for politicians and planners as they think about transportation (and energy and climate) policy.

Let's be clear: there is no alternative to pooling. Without pooling, masses of vehicles will be stuffed onto our streets and roads. If all are electric, yes, the climate impacts are modest, which is good (assuming the trend toward energy decarbonization continues). And yes, if all vehicles are automated and the entire road system is converted to handle AVs (with dedicated, narrow lanes), the large increase in vehicle use could possibly be accommodated with existing road infrastructure. But with so much more vehicle use, much of the benefit of electrification would be eroded, urban sprawl would be accelerated, and the gap between haves and have-nots would widen. Richer people would travel in more expensive AVs, and the poor and physically limited would continue to be marginalized. The result would be the nightmare scenario of chapter 1.

Most travelers will not easily embrace pooling or AVs. Most transit operators and their advocates will continue to feel threatened by ride-hailing companies. Trade unions for professional drivers will stand in the way. Automakers and related industries such as parts suppliers will stick to internal combustion engines and partial automation as long as they can. The oil industry will throw its weight behind campaigns to maintain the status quo as long as possible. Politicians will shy away from pricing reforms and embracing private mobility companies. The transition will feel slow to those of us eager to create more sustainable transportation. All this is assured.

But momentum in the direction of pooling is gathering, and at some point, it will become irreversible. There is too much at stake for business as usual to go on indefinitely. As electric and AV technology comes along, the push toward pooling becomes even more compelling and consequential. It is a future that we must steer toward if we want a better world.

Notes

Chapter 1

1. Paul Barter, "'Cars Are Parked 95% of the Time.' Let's Check!," Reinventing Parking, 22 February 2013.

2. US Department of Transportation, National Transportation Statistics, Table 1-40, updated April 2017.

3. These figures are based on fifteen thousand miles of travel per year. AAA, "Your Driving Costs," AAA NewsRoom, 23 August 2017; US Department of Labor, 2015 Consumer Expenditure Survey, Table 1203.

4. National Safety Council, "NSC Motor Vehicle Fatality Estimates," 2017.

5. US Energy Information Administration, "Petroleum and Other Liquids, U.S. Weekly Product Supplied," accessed 3 September 2017.

6. US Energy Information Administration, "U.S. Energy-Related Carbon Dioxide Emissions, 2015," March 2017.

7. "Singapore: Intelligent Transport System," C40 Cities, 2013; Brian Fung, "Here's One Place in the World You Can Already Hail a Driverless Taxi," *Washington Post*, 25 August 2016.

8. Jha Alok, "Delhi's Air Pollution Is Causing a Health Crisis: So, What Can Be Done?" *The Guardian*, 3 November 2015.

9. Sharon Silke Carty, "How Humans Could Ruin the Autonomous Era," *Automotive News*, 24 October 2016.

10. Chinese sales of new-energy vehicles in 2016 totaled 507,000, consisting of 409,000 all-electric vehicles and 98,000 plug-in hybrid vehicles (many of these trucks and buses), according to Liu Wanxiang, "中汽协: 2016年新能源汽车产销量均超50万辆,同比增速约50%" [China Auto Association: 2016 New Energy Vehicle Production and Sales Were over 500,000, an Increase of About 50%], D1EV, accessed 12 January 2017.

11. International Energy Agency, *Global EV Outlook 2017* (Paris, France: OECD/IEA, 2017).

12. European Alternative Fuels Observatory, "Norway," accessed 4 September 2017.

13. John Zimmer, "The Third Transportation Revolution: Lyft's Vision for the Next Ten Years and Beyond," Medium, 18 September 2016.

14. The cost per mile for an electric automated vehicle operated fifty thousand miles or more per year was calculated to be about the same as for a gasoline-powered AV, with the higher up-front costs offset by the lower operating costs (for maintenance and energy). See following note.

15. These numbers are based on a cost analysis of automated and pooled cars done by Aditi Meshram and Daniel Sperling of the Institute of Transportation Studies, University of California, Davis, based on numbers from the Internal Revenue Service, the BLS Consumer Expenditure Survey, APTA's *2016 Public Transportation Fact Book*, and unpublished sources. For transit costs, see chapter 5.

16. T. S. Stephens et al., "Estimated Bounds and Important Factors for Fuel Use and Consumer Costs of Connected and Automated Vehicles," US Department of Energy, National Renewable Energy Laboratory, Technical Report NREL/TP-5400-67216, November 2016.

17. José Viegas, Luis Martinez, and Philippe Crist, "Shared Mobility: Innovation for Liveable Cities," International Transport Forum Corporate Partnership Board, 2016.

18. Zia Wadud, Don Mackenzie, and Paul N. Leiby, "Help or Hindrance? The Travel, Energy, and Carbon Impacts of Highly Automated Vehicles," *Transportation Research Part A* 86 (April 2016): 1–18.

19. Daniel Fagnant and Kara Kockelman, "The Travel and Environmental Implications of Shared Autonomous Vehicles, Using Agent-Based Model Scenarios," *Transportation Research Part C* 40 (2014): 1–13.

20. Lew Fulton, Jacob Mason, and Dominique Meroux, "Three Revolutions in Urban Transportation," Institute of Transportation Studies, University of California, Davis, Research Report UCD-ITS-RR-17-03, 2017.

21. Brian D. Taylor and Rebecca Kalauskas, "Addressing Equity in Political Debates over Road Pricing: Lessons from Recent Projects," *Transportation Research Record* 2187 (2010): 44–52.

Chapter 2

1. Chinese sales of new-energy vehicles in 2016 totaled 507,000, consisting of 409,000 all-electric vehicles and 98,000 plug-in hybrid vehicles (many of these trucks and buses), according to Liu Wanxiang, "中汽协: 2016年新能源汽车产销量均超50万辆,同比增速约50%" [China Auto Association: 2016 New Energy Vehicle Production and Sales Were over 500,000, an Increase of About 50%], D1EV, accessed 12 January 2017.

2. Surveys and interviews were conducted in eight states. The two principal reports are Kenneth Kurani et al., "New Car Buyers' Valuation of Zero-Emission Vehicles: California," Institute of Transportation Studies, University of California, Davis, Research Report UCD-ITS-RR-16-05, 2016; and Kenneth Kurani and Nicolette Caperello, "New Car Buyers' Valuation of Zero-Emission Vehicles: Northeast States for Coordinated Air Use Management (NESCAUM)," Institute of Transportation Studies, University of California, Davis, Research Report UCD-ITS-RR-16-16, 2016.

3. Bertel Schmitt, "There Is a Lot of EV Talk in Europe. And Very Little Buying," *Forbes*, 6 October 2016.

4. Dana Hull, "Tesla Deliveries Miss Forecasts Again on Production Delays," Bloomberg Markets, 3 January 2017.

5. Yunshi Wang et al., "China's Electric Car Surge," *Energy Policy* 102 (March 2017): 486–90.

6. R. Sims et al., "Transport," in *Climate Change 2014: Mitigation of Climate Change; Contribution of Working Group III to the Fifth Assessment Report of the Intergovernmental Panel on Climate Change*, ed. O. Edenhofer et al. (Cambridge: Cambridge University Press, 2014), 599–670.

7. Damian Carrington, "Electric Cars 'Will Be Cheaper than Conventional Vehicles by 2022,'" *The Guardian*, 25 February 2016.

8. Craig Morris, "India Joins Norway and Netherlands in Wanting 100% Electric Vehicles," RenewEconomy, 4 April 2016; PTI, "India Looks to Put Hybrid and Electric Vehicles on the Fast Track in 2016," *The Times of India*, 29 December 2015.

9. Bertel Schmitt, "Germany's Bundesrat Resolves End of Internal Combustion Engine," *Forbes*, 8 October 2016.

10. Mark Singer, "Consumer Views on Plug-In Electric Vehicles—National Benchmark Report," National Renewable Energy Laboratory, Technical Report NREL/TP-5400-65279, January 2016.

11. International Energy Agency, "A Brief History of Electric Vehicles," in *Global EV Outlook: Understanding the Electric Vehicle Landscape to 2020* (Paris, France: IEA, April 2013); "Electric Vehicles History Part II: Early History," Electric Vehicles News, accessed 3 September 2017.

12. The following historical review is from Daniel Sperling, *Future Drive* (Washington, DC: Island Press, 1995), 36–37.

13. This discussion of CARB and the early years of the ZEV mandate borrows heavily from "California's Pioneering Role," in *Two Billion Cars*, by Daniel Sperling and Deborah Gordon (Oxford: Oxford University Press, 2009), ch. 7; and Gustavo Collantes and Daniel Sperling, "The Origin of California's Zero Emission Vehicle Mandate," *Transportation Research Part A* 42 (2008): 1302–13.

14. The European Union has been the world leader in enacting climate policy. It adopted (voluntary) carbon dioxide standards for vehicles in 1998 and launched a cap-and-trade program for major stationary sources in 2005 and various other measures after that. California's 2006 global warming law is broader, requiring reductions across the entire economy, but those rules weren't enacted until 2010 and beyond. Also, as a state within a nation, California has limited jurisdiction over ocean shipping, aviation, and other activities that cross state borders.

15. A later version of the EV1 was outfitted with nickel-metal-hydride batteries that provided 26 kWh of capacity and doubled the range, but only about four hundred cars were produced with these batteries.

16. "General Motors EV1," Wikipedia entry, last modified 23 July 2017.

17. See Levi Tillemann, *The Great Race: The Global Quest for the Car of the Future* (New York: Simon & Schuster, 2015).

18. Max Fawcett, "Is Tesla's Model-S the Beginning of the End for Oil?," *Alberta Oil*, 2 July 2015.

19. Keith Naughton, "Bob Lutz: The Man Who Revived the Electric Car," *Newsweek*, 22 December 2007.

20. Fawcett, "Tesla's Model-S."

21. David Reichmuth and Don Anair, "Electrifying the Vehicle Market: Evaluating Automaker Leaders and Laggards in the United States," Union of Concerned Scientists, August 2016.

22. "The Week That Electric Vehicles Went Mainstream," Tesla website, 7 April 2016.

23. The actual rule is far more complicated. It has two main categories: pure zero-emission vehicles (ZEVs) and transitional zero-emission vehicles (TZEVs), which include plug-in hybrid EVs. Clean conventional vehicles were credited in the past, but the categories have been simplified. In practice, a complicated credit multiplier system was devised that results in a much larger number of TZEVs being allowed. At the March 2008 board meeting, a resolution I put forward was adopted to redesign the entire program in 2009 around the original goal of accelerating the commercialization of fuel cell, battery, and plug-in hybrid technologies—this time motivated more by climate and energy goals than by local air pollution concerns. An ARB staff ZEV mandate tutorial can be found here: https://www.arb.ca.gov/msprog/zevprog/zevtutorial/zevtutorial.htm.

24. Keith Naughton, John Lippert, and Jamie Butters, "Ford Willing to Work with Trump If Policies Are Right," *Detroit News*, 3 December 2016.

25. National Research Council, *Transitions to Alternative Vehicles and Fuels* (Washington, DC: National Academies Press, 2013).

26. Elisabeth Behrmann, "BMW Sees Battery Costs Weighing on EVs," *Automotive News Europe*, 5 December 2016.

27. Carrington, "Electric Cars 'Will Be Cheaper.'"

28. Peter Slowik and Nic Lutsey, "Evolution of Incentives to Sustain the Transition to a Global Electric Vehicle Fleet," International Council on Clean Transportation (ICCT) white paper, November 2016; Paul Wolfram and Nic Lutsey, "Electric Vehicles: Literature Review of Technology Costs and Carbon Emissions," ICCT white paper, July 2016.

29. Joan M. Ogden, Lewis Fulton, and Daniel Sperling, "Making the Transition to Light-Duty Electric-Drive Vehicles in the U.S.: Costs in Perspective to 2035," Institute of Transportation Studies, University of California, Davis, Research Report UCD-ITS-RR-16-21, 2016.

30. Ibid.

31. Fawcett, "Tesla's Model-S."

32. David Coady et al., "How Large Are Global Energy Subsidies?," International Monetary Fund, 18 May 2015.

33. Scott Hardman et al., "The Effectiveness of Financial Purchase Incentives for Battery Electric Vehicles: A Review of the Evidence," *Renewable and Sustainable Energy Reviews* 80 (December 2017): 1100–1111.

34. Wang et al., "China's Electric Car Surge."

35. Norwegian Electric Vehicle Association, "Norwegian EV Market," last updated 30 June 2017.

36. European Alternative Fuels Observatory, "Norway," accessed 4 September 2017.

37. David Jolly, "Norway Is a Model for Encouraging Electric Car Sales," *New York Times*, 16 October 2015.

38. Erik Figenbaum, "Perspectives on Norway's Supercharged Electric Vehicle Policy," *Environmental Innovation and Societal Transitions*, 17 November 2016; Kristin Ystmark Bjerkan, Tom E. Nørbech, and Marianne Elvsaas Nordtømm, "Incentives for Promoting Battery Electric Vehicle (BEV) Adoption in Norway," *Transportation Research Part D* 43 (2016): 169–80.

39. Norwegian Electric Vehicle Association, "Norwegian EV Policy," last updated 30 June 2017.

40. Lingzhi Jin, Stephanie Searle, and Nic Lutsey, "Evaluation of State-Level U.S. Electric Vehicle Incentives," ICCT white paper, October 2014.

41. Suzanne Guinn, "EVSE Rebates and Tax Credits, by State," clippercreek .com, accessed 22 December 2016.

42. James Murray, "Electric Vehicle Charge Points to Outnumber Petrol Stations by 2020, Say Nissan," *The Guardian*, 4 August 2016.

43. Paul McVeigh, "Ford, BMW, Daimler, VW Group Plan Fast-Charging EV Network," *Automotive News Europe*, 29 November 2016.

44. Zach McDonald, "A Simple Guide to DC Fast Charging," fleetcarma .com, 4 February 2016.

Chapter 3

1. Nelson Chan and Susan Shaheen, "Ridesharing in North America: Past, Present, and Future," *Transport Reviews* 32 (January 2012): 93–112.

2. US Census Bureau, 2014 American Community Survey 1-Year Estimates, Table B08006.

3. John Zimmer, "The Third Transportation Revolution: Lyft's Vision for the Next Ten Years and Beyond," Medium, 18 September 2016.

4. Eric Newcomer, "GM Invests $500 Million in Lyft," Bloomberg Technology, 4 January 2016.

5. Nellie Bowles and Danny Yadron, "Self-Driving Cars Hog the Road at CES," *The Guardian*, 7 January 2016.

6. Zimmer, "Third Transportation Revolution."

7. Faiz Siddiqui, "Uber Is Betting D.C. Commuters Are Willing to Pay to Slug," *Washington Post*, 27 March 2017.

8. Meg Graham, "Uber Tests Commuting Service for Drivers with a Seat to Spare," *Chicago Tribune*, 8 December 2015.

9. Manish Singh, "Uber Eyes UberCommute Expansion in India," Mashable, 24 November 2016.

10. Susan Shaheen, Adam Stocker, and Marie Mundler, "Online and App-Based Carpooling in France: Analyzing Users and Practices—A Study of BlaBlaCar," in *Disrupting Mobility: Impacts of Sharing Economy and Innovative Transportation on Cities*, ed. Gereon Meyer and Susan Shaheen (New York: Springer, 2017), 181–96.

11. "BlaBlaCar Unveils Opel Leasing Deal in Boost for Ride-Sharing," *Automotive News*, 6 April 2017.

12. Susan Shaheen, Nelson D. Chan, and Teresa Gaynor, "Casual Carpooling in the San Francisco Bay Area: Understanding Characteristics, Behaviors, and Motivations," *Transport Policy* 51 (January 2016): 165–73.

13. Ibid.

14. Lisa Rayle et al., "App-Based, On-Demand Ride Services: Comparing Taxi and Ridesourcing Trips and User Characteristics in San Francisco," Institute of Transportation Studies, University of California, Davis, Research Report UCTC-FR-2014-08, August 2014.

15. The share of travelers using carpooling for commuting in the Bay Area has remained at about 10 percent over the last decade, according to the US Census Bureau's American Community Survey, accessed 10 April 2017.

16. "Transit Ridership," Vital Signs, updated December 2015.

17. Adam Cohen and Susan Shaheen, "Planning for Shared Mobility," American Planning Association, Planning Advisory Service Report 583, 2016.

18. Barbara Laurenson, personal communication, 1 February 2017.

19. "Announcing UberPool," Uber Newsroom, 5 August 2014.

20. Oliver Smith, "Uber Taxi App: Founder Travis Kalanick's Plan to Rid London of a Million Cars," *City A.M.*, 6 October 2014.

21. José Viegas, Luis Martinez, and Philippe Crist, "Shared Mobility: Innovation for Liveable Cities," International Transport Forum Corporate Partnership Board, 2016.

22. Javier Alonso-Mora et al., "On-Demand High-Capacity Ride-Sharing via Dynamic Trip-Vehicle Assignment," *Proceedings of the National Academy of Sciences* 114 (17 January 2017): 462–67.

23. Jeffrey B. Greenblatt and Samveg Saxena, "Autonomous Taxis Could Greatly Reduce Greenhouse Gas Emissions of U.S. Light-Duty Vehicles," *Nature Climate Change* 5 (2015): 860–63.

24. Bruce Schaller, "Unsustainable? The Growth of App-Based Ride Services and Traffic, Travel and the Future of New York City," Schaller Consulting, 27 February 2017.

25. Aaron Sankin, "Can a Smartphone App Save the Taxi Industry from Uber?," The Daily Dot, 9 May 2016.

26. Joe Fitzgerald Rodriguez, "Flywheel Taxi App Sold in Effort to Battle Uber, Lyft," *San Francisco Examiner*, 10 April 2017.

27. "Verifone Launches Curb in Five New U.S. Cities," Verifone press release, 21 July 2016.

28. David Mahfouda, personal communication, 30 January 2017.

29. Robert Cervero, *Paratransit in America: Redefining Mass Transportation* (Westport, CT: Praeger, 1997).

30. James Covert, "Taxi-Sharing App Doesn't Want You Waiting on Line at the Airport," *New York Post*, 22 May 2015.

31. Janelle Orsi, "Taxi Cab Sharing in New York City and Beyond," Shareable, 23 February 2010.

32. Elizabeth Daigneau, "New Yorkers 'Share a Cab,'" Governing, 4 March 2010.

33. Paolo Santi et al., "Quantifying the Benefits of Vehicle Pooling with Shareability Networks," *Proceedings of the National Academy of Sciences* 111 (16 September 2014): 13290–94.

34. Laura Bliss, "What's behind Declining Transit Ridership Nationwide," CityLab, 24 February 2017.

35. Lisa Rayle et al., "Just a Better Taxi? A Survey-Based Comparison of Taxis, Transit, and Ridesourcing Services in San Francisco," *Transport Policy* 45 (January 2016): 168–78.

36. Matthew Daus, "When the L Closes Let Dollar Vans Rush In," *New York Daily News*, 10 October 2016.

37. David A. King and Eric Goldwyn, "Why Do Regulated Jitney Services Often Fail? Evidence from the New York City Group Ride Vehicle Project," *Transport Policy* 35 (September 2014): 186–92.

38. David Holmes, "Anti-Uber: The Quiet Disruption of NYC Dollar Vans," Pando, 8 July 2014.

39. "Current Licensees," New York Taxi and Limousine Commission, last updated 2 August 2017.

40. Matt Daus, personal communication, 30 March 2017.

41. Cohen and Shaheen, "Planning for Shared Mobility."

42. Linda Poon, "Bridj Collapses after Just 3 Years," CityLab, 1 May 2017.

43. Lew Pratsch, "Commuter Ridesharing," in *Public Transportation: Planning, Operations, and Management*, ed. George E. Gray and Lester A. Hoel (Englewood Cliffs, NJ: Prentice Hall, 1979), 168–87.

44. Cathy Yang Liu and Gary Painter, "Travel Behavior among Latino Immigrants: The Role of Ethnic Neighborhoods and Ethnic Employment," *Journal of Planning Education* 32 (January 2012): 62–80.

45. Shaheen, Chan, and Gaynor, "Casual Carpooling."

46. Ibid.

47. Marc Oliphant, "The Native Slugs of Northern Virginia: A Profile of Slugging in the Washington DC Region," master's thesis, Virginia Polytechnic Institute, Blacksburg, 2008; Vanasse Hangen Brustlin Inc., "Dynamic Ridesharing (Slugging) Data," final report prepared for Virginia Department of Transportation, 15 June 2006.

48. Biz Carson, "Uber Responds: Google Claims about Stolen Technology Are a Total 'Misfire,'" Business Insider, 7 April 2017.

49. Ibid.

50. Adam Stocker and Susan Shaheen, "Shared Automated Vehicles: Review of Business Models," presentation for the Roundtable on Cooperative Mobility Systems and Automated Driving, International Transport Forum, 6 December 2016.

51. Alex Davies, "Turns Out the Hardware in Self-Driving Cars Is Pretty Cheap," Wired, 22 April 2015; Xavier Mosquet et al., "Revolution in the Driver's Seat: The Road to Autonomous Vehicles," bcg.perspectives, 21 April 2015.

52. Newcomer, "GM Invests $500 Million."

53. Zimmer, "Third Transportation Revolution."

54. Dana Hull, "Elon Musk Says Tesla Car-Share Network Is 'the People vs. Uber,'" Bloomberg Technology, 26 October 2016.

55. Kirsten Korosec, "Why Europe's Biggest Railway Is Working on Self-Driving Cars," *Fortune*, 7 May 2016.

56. US Department of Transportation, "Smart City Challenge: List of Applicants," 2016.
57. San Francisco Municipal Transportation Agency, "City of San Francisco: Meeting the Smart City Challenge," sfmta.com, 2016.
58. Donald Shoup, *The High Cost of Free Parking* (Chicago: APA Planners, 2005).
59. US Department of Transportation, NHTSA, "Federal Automated Vehicles Policy," September 2016.

Chapter 4

1. Ryan Bradley, "Tesla Autopilot: The Electric-Vehicle Maker Sent Its Cars a Software Update That Suddenly Made Autonomous Driving a Reality," *MIT Technology Review*, March/April 2016.
2. "Autonomous Cars: Self-Driving the New Auto Industry Paradigm," Morgan Stanley Research, November 2013.
3. Emily Barasch, "Study: Nearly All Cars to Be Self-Driving by 2050," mic.com, 5 January 2014.
4. Cadie Thompson, "Elon Musk Says Tesla's Fully Autonomous Cars Will Hit the Road in 3 Years," Business Insider, 25 September 2015; "All Tesla Cars Being Produced Now Have Full Self-Driving Hardware," Tesla Blog, 19 October 2016.
5. "Ford Targets Fully Autonomous Vehicle for Ride Sharing in 2021; Invests in New Tech Companies, Doubles Silicon Valley Team," Ford press release, 16 August 2016.
6. "Forecasts," Driverless Car Market Watch, accessed 4 September 2017.
7. Alex Roy, "What If the Autonomous Car Industry Is Wrong?," The Drive, 7 December 2016.
8. Chris Urmson, "Google Self-Driving Car Project," SXSW Interactive Featured Session, 2016.
9. "Gartner's 2016 Hype Cycle for Emerging Technologies Identifies Three Key Trends That Organizations Must Track to Gain Competitive Advantage," Gartner Newsroom press release, 16 August 2016.
10. See the crash fatalities estimate from the World Health Organization, "Global Status Report on Road Safety 2013," available at http://www.who.int/iris/bitstream/10665/78256/1/9789241564564_eng.pdf.
11. "Self-Driving Cars: The Next Revolution," KPMG and the Center for Automotive Research white paper, 28 November 2012.

12. Jonas Meyer et al., "Autonomous Vehicles: The Next Jump in Accessibilities?," *Research in Transportation Economics* 62 (June 2017): 80–91.
13. Dimitris Milakis, Bart van Arem, and Bert van Wee, "Policy and Society Related Implications of Automated Driving: A Review of Literature and Directions for Future Research," *Journal of Intelligent Transportation Systems* 21, no. 4, published online 13 February 2017.
14. "Wasted Spaces: Options to Reform Parking Policy in Los Angeles," Council of Infill Builders, 16 May 2017.
15. Will Knight, "10-4, Good Computer: Automated System Lets Trucks Convoy as One," *MIT Technology Review*, 28 May 2014; Xiao-Yun Lu and Steven E. Shladover, "Automated Truck Platoon Control and Field Test," in *Road Vehicle Automation*, ed. Gereon Meyer and Sven Beiker (Switzerland: Springer International, 2014), 247–61.
16. Tom Simonite, "Uber Is Betting We'll See Driverless 18-Wheelers before Taxis," *MIT Technology Review*, 7 September 2016.
17. Greg Harman, "Driverless Big Rigs: New Technologies Aim to Make Trucking Greener and Safer," *The Guardian*, 24 February 2015.
18. "Automated Driving," SAE International, 2014; "Taxonomy and Definitions for Terms Related to Driving Automation Systems for On-Road Motor Vehicles," SAE International, 30 September 2016.
19. Patrice Reilhac et al., "User Experience with Increasing Levels of Vehicle Automation: Overview of the Challenges and Opportunities as Vehicles Progress from Partial to High Automation," in *Automotive User Interfaces*, ed. Gerrit Meixner and Christian Muller (Switzerland: Springer International, 2017), 457–82.
20. Keith Naughton, "Ford's Dozing Engineers Side with Google in Full Autonomy Push," *Automotive News*, 17 February 2017; Aaron Birch, "Ford Denies Report That Engineers Are Dozing in Self-Driving Test Cars," Ford Authority newsletter, 20 February 2017.
21. Naughton, "Ford's Dozing Engineers."
22. Fred Lambert, "Elon Musk Defends Level 3 Autonomy against Google and Volvo, Says 'Morally Wrong to Withhold Functionalities That Improve Safety,'" Electrek, 13 September 2016.
23. John R. Quain, "Makers of Self-Driving Cars Ask What to Do with Human Nature," *New York Times*, 7 July 2016.
24. Naughton, "Ford's Dozing Engineers."

25. Bradley Berman, "Whoever Owns the Maps Owns the Future of Self-Driving Cars," *Popular Mechanics*, 1 July 2016.

26. Carolyn Said, "Now at Udacity, Google X Founder Talks Self-Driving Cars and Jobs," *San Francisco Chronicle*, 3 December 2016.

27. Kathy Pretz, "The Drivers behind Autonomous Vehicles," The Institute, 7 April 2014.

28. Steven E. Schladover, "PATH at 20: History and Major Milestones," *IEEE Transactions on Intelligent Transportation Systems* 8 (December 2007): 584–92.

29. No Hands Across America website, http://www.cs.cmu.edu/~tjochem/nhaa/nhaa_home_page.html.

30. Alex Oagana, "A Short History of Mercedes-Benz Autonomous Driving Technology," autoevolution.com, 26 January 2016.

31. DARPA, "The DARPA Grand Challenge: Ten Years Later," 13 March 2014.

32. "Dutch Business Community Welcomes Truck Platoons," European Truck Platooning website, press releases, accessed 4 September 2017; James Vincent, "Self-Driving Truck Convoy Completes Its First Major Journey across Europe," The Verge, 7 April 2016.

33. Hope Reese, "US DOT Unveils 'World's First Autonomous Vehicle Policy,' Ushering in Age of Driverless Cars," TechRepublic, 20 September 2016; US Department of Transportation, NHTSA, "Federal Automated Vehicles Policy," September 2016.

34. Adam Lashinsky, *Inside: Uber's Quest for World Domination* (New York: Penguin, 2017); Brad Stone, *The Upstarts: How Uber, Airbnb, and the Killer Companies of the New Silicon Valley Are Changing the World* (New York: Little, Brown, 2017).

35. Josh Lowensohn, "Uber Gutted Carnegie Mellon's Top Robotics Lab to Build Self-Driving Cars," The Verge, 19 May 2015.

36. Bradley, "Tesla Autopilot"; "Autopilot," Tesla Press Information, n.d., https://www.tesla.com/presskit/autopilot#autopilot.

37. Darrell Etherington, "Comma.ai Cancels the Comma One Following NHTSA Letter," TechCrunch, 28 October 2016.

38. David Shepardson, "Ford, Volvo, Google, Uber, Lyft Form Coalition to Promote Self-Driving Cars," *Automotive News*, 26 April 2016.

39. Hod Lipson and Melba Kurman, *Driverless: Intelligent Cars and the Road Ahead* (Cambridge, MA: MIT Press, 2016).

40. Todd Littman, "Autonomous Vehicle Implementation Predictions," Victoria Transport Policy Institute, 27 February 2017.

41. Brandon Schoettle and Michael Sivak, "Motorists' Preferences for Different Levels of Vehicle Automation," University of Michigan Transportation Research Institute, Report UMTRI-2015-22, July 2015.

42. Insurance Information Institute, "Self-Driving Cars and Insurance," July 2016.

43. Boston Consulting Group, "Self-Driving Vehicles, Robo-Taxis, and the Urban Mobility Revolution," bcg.perspectives, 21 July 2016.

44. Department of Statistics Singapore, "Latest Data," accessed 4 September 2017.

45. "Roadways in Singapore," Trade Chakra, 2008; "Singapore: Intelligent Transport System," C40 Cities, 2013.

46. K. Spieser et al., "Toward a Systematic Approach to the Design and Evaluation of Automated Mobility-on-Demand Systems: A Case Study in Singapore," in *Road Vehicle Automation*, ed. Gereon Meyer and Sven Beiker (Switzerland: Springer International, 2014), 229–45.

47. Singapore Government Land Transport Authority, "Electronic Road Pricing (ERP)," accessed 4 September 2017.

48. Singapore Government Ministry of Transport, "Driverless Vehicles: A Vision for Singapore's Transport," accessed 4 September 2017.

49. Jon Russell, "MIT Spinout NuTonomy Just Beat Uber to Launch the World's First Self-Driving Taxi," TechCrunch, 24 August 2016; Andrew J. Hawkins, "Singapore's Self-Driving Cars Can Now Be Hailed with a Smartphone," The Verge, 22 September 2016.

50. Adrian Lim and Chew Hui Min, "nuTonomy Resumes Driverless Car Trials in One-North, Says Software Glitch to Blame for Accident," Straits Times, 24 November 2016.

51. James Hedlund, "Autonomous Vehicles Meet Human Drivers: Traffic Safety Issues for States," Governors Highway Safety Association, 2016.

52. Carrie Cox and Andrew Hart, "How Autonomous Vehicles Could Relieve or Worsen Traffic Congestion," HERE white paper, 2016.

53. Kenneth P. Laberteaux et al., "Methodology for Gauging Usage Opportunities for Partially Automated Vehicles with Application to Public Travel Survey Data Sets," *Transportation Research Record* 2625 (2017): 43–50.

54. Roy, "What If?"

55. Littman, "Autonomous Vehicle"; Corey D. Harper et al., "Estimating Potential Increases in Travel with Autonomous Vehicles for the Non-driving, Elderly and People with Travel-Restrictive Medical Conditions," *Transportation Research Part C* 72 (November 2016): 1–9.

56. Ariel Wittenberg, "Helping to Build 'Moral Machines' at MIT," Greenwire, 18 October 2016.

57. Michael Taylor, "Self-Driving Mercedes-Benzes Will Prioritize Occupant Safety over Pedestrians," *Car and Driver*, 7 October 2016.

58. Jean-Francois Bonnefon, Azim Shariff, and Iyad Rahwan, "The Social Dilemma of Autonomous Vehicles," *Science*, 24 June 2016, 1573.

59. Bryan Casey, "Amoral Machines or How Roboticists Can Learn to Stop Worrying and Love the Law," *Northwestern University Law Review* 111, no. 5 (2017): n.p.

60. Andy Greenberg, "Hackers Remotely Kill a Jeep on the Highway—with Me in It," Wired, 21 July 2015.

61. Alexandria Sage, "CIA 'Mission' on Cars Shows Concern about Next-Generation Vehicles," *Automotive News*, 9 March 2017.

62. See Erik Brynjolfsson and Andrew McAfee, *The Second Machine Age: Work, Progress, and Prosperity in a Time of Brilliant Technologies* (New York: Norton, 2016).

63. Yoni Heilbronn, chief marketing officer for Argus Cyber Security, quoted in Andy Kulisch, "Cybersecurity Push May Tie Up Autonomous-Car Legislation," *Automotive News*, 24 June 2017.

64. Sage, "CIA 'Mission.'"

65. See chapter 5 in this book. Bus and rail transit in the United States account for about 1 percent of passenger miles, with taxis and ridehailing services representing much less than that.

66. Said, "Now at Udacity."

67. International Transportation Forum, "Managing the Transition to Driverless Road Freight Transport," OECD, 2017.

68. Ibid.

69. Prateek Bansal and Kara M. Kockelman, "Forecasting Americans' Long-Term Adoption of Connected and Autonomous Vehicle Technologies," *Transportation Research Part A* 95 (January 2017): 49–63.

70. James M. Anderson et al., "Autonomous Vehicle Technology: A Guide for Policymakers," RAND Corporation Research Report RR-443-2-RC,

2016; Daniel Fagnant and Kara Kockelman, "Preparing a Nation for Autonomous Vehicles: Opportunities, Barriers and Policy Recommendations," *Transportation Research Part A* 77 (2015): 167–81.

71. US Department of Transportation, NHTSA, "Federal Automated Vehicles Policy," September 2016.
72. An extensive review of the benefits and costs of automated vehicles that focuses specifically on policy implications for Minnesota suggests that if Minnesota were to adopt statutes regarding testing of AVs within the state, it could become a testing ground for innovations that address issues AVs have with driving in snow-covered landscapes. Adeel Lari, Frank Douma, and Ify Onyiah, "Self-Driving Vehicles and Policy Implications: Current Status of Autonomous Vehicle Development and Minnesota Policy Implications," *Minnesota Journal of Law, Science & Technology* 16 (2015): 735–68.
73. Robin Chase, "Shared Passenger Mobility Protocol for Livable Cities," OSMOSYS, accessed 4 September 2017. See also National Association of City Transportation Officials, "NACTO Policy Statement on Automated Vehicles," 22 June 2016.

Chapter 5

1. Adella Santos et al., *Summary of Travel Trends: 2009 National Household Travel Survey* (Washington, DC: Federal Highway Administration, 2011), Table 9.
2. US Department of Transportation, National Transportation Statistics, Table 1-40, updated April 2017.
3. Michael Manville, David A. King, and Michael J. Smart, "The Driving Downturn: A Preliminary Assessment," *Journal of the American Planning Association* 83 (2017): 49.
4. Robert Poole, "Autonomous Vehicles' Disruptive Potential for Transit," *Surface Transportation News*, 7 January 2016.
5. Chris Martin and Joe Ryan, "Super-Cheap Driverless Cabs to Kick Mass Transit to the Curb," Bloomberg Technology, 24 October 2016.
6. Henry Grabar, "Is Uber Killing the Public Bus, or Helping It?," Slate, 12 September 2016.
7. Sharon Feigon and Colin Murphy, "Shared Mobility and the Transformation of Public Transit," Transportation Research Board, TCRP Research Report 188, 2016.

8. John Neff and Matthew Dickens, *2015 Public Transportation Fact Book, Appendix A: Historical Tables* (Washington, DC: American Public Transportation Association, June 2015), Tables 1 and 8.

9. David W. Jones, *Urban Transit Policy: An Economic and Political History* (Englewood Cliffs, NJ: Prentice Hall, 1985).

10. Steven Polzin and Alan E. Pisarski, "Brief 10: Commuting Mode Choice," *Commuting in America 2013* (Washington, DC: AASHTO, October 2013).

11. John Neff and Matthew Dickens, *2012 Public Transportation Fact Book, Appendix A: Historical Tables* (Washington, DC: American Public Transportation Association, March 2012), Table 71.

12. In 2013, the average fare for both urban and suburban bus and rail transit in the United States was $0.28 per mile. The total operating plus capital cost for all metropolitan transit service was $1.13 per mile ($0.78 for operating costs plus $0.35 for capital costs). For buses (a category that includes aerial tram, bus, bus rapid transit, commuter bus, demand-response paratransit, ferry, jitney, trolley bus, and vanpool, with buses representing all but a tiny percent of the category), the costs were $1.36 per mile ($1.12 for operating costs plus $0.24 for capital costs). These costs per mile varied greatly from one location to another and by time of day, mostly related to ridership. Office of Budget and Policy, *2015 National Transit Summary and Trends* (Washington, DC: US Department of Transportation, Federal Transit Administration, October 2016); John Neff and Matthew Dickens, *2015 Public Transportation Fact Book* (Washington, DC: American Public Transportation Association, November 2015).

13. Manville, King, and Smart, "Driving Downturn."

14. Office of Budget and Policy, *Transit Profiles: 2015 Report Year Summary* (Washington, DC: US Department of Transportation, Federal Transit Administration, September 2016).

15. Lisa Rayle et al., "Just a Better Taxi? A Survey-Based Comparison of Taxis, Transit, and Ridesourcing Services in San Francisco," *Transport Policy* 45 (January 2016): 168–78; Feigon and Murphy, "Shared Mobility"; Bruce Schaller, "Unsustainable? The Growth of App-Based Ride Services and Traffic, Travel and the Future of New York City," Schaller Consulting, 27 February 2017.

16. Promoted by transit agencies such as MARTA: http://www.itsmarta.com/tod-overview.aspx.

17. Bill Conerly, "Self-Driving Cars Will Kill Transit-Oriented Development," *Forbes*, 8 August 2016.
18. American Public Transportation Association, "Public Transportation Benefits," 2017.
19. In 2012, cars used 3,193 Btu of energy per passenger mile compared to 4,030 for urban transit buses. Light-duty trucks used 6,674 Btu per vehicle mile. Assuming the light trucks have the same average vehicle occupancy as cars, their energy use would be slightly higher than buses per passenger mile. Stacy C. Davis, Susan W. Diegel, and Robert G. Boundy, *Transportation Energy Data Book*, 33rd ed. (Washington, DC: US Department of Energy, July 2014), Table 2.13.
20. Ibid.
21. Melvin M. Webber, "The Marriage of Transit and Autos: How to Make Transit Popular Again," *Access* 5 (Fall 1994): 26–31.
22. Jeff Spross, "Why Replacing the Bus with Uber Is Actually Pretty Smart," *The Week*, 16 August 2016.
23. "Autonomous Vehicles: A Potential Game Changer for Urban Mobility," Combined Mobility Platform, International Association of Public Transport (UITP) policy brief, January 2017; Emily Badger, "What Will Happen to Public Transit in a World Full of Autonomous Cars?," CityLab, 17 January 2014; Carlo Sessa et al., "Results of the On-line DELPHI Survey," CityMobil2, European Union Seventh Framework Programme, 2015.
24. Joshua Brustein, "Uber and Lyft Want to Replace Public Buses," Bloomberg Technology, 15 August 2016.
25. Craig S. Smith, "A Canadian Town Wanted a Transit System. It Hired Uber," *New York Times*, 16 May 2017.
26. Spencer Woodman, "Welcome to Uberville," The Verge, 2 September 2016; Eric Jaffe, "Uber and Public Transit Are Trying to Get Along," CityLab, 3 August 2015.
27. Robert A. Cronkleton, "Kansas City's Microtransit Experiment RideKC: Bridj Launches Monday," *Kansas City Star*, 3 March 2016.
28. "A for Effort, but So Far, a Bridj to Nowhere," Transit Center, 24 February 2017.
29. Susan Shaheen et al., "RideKC: Bridj Pilot Evaluation: Impact, Operational, and Institutional Analysis," Transportation Sustainability Research Center, UC Berkeley, October 2016.

30. Internal memorandum prepared for an LA Metro advisory board, April 2017.

31. Joseph Kane, Adie Tomer, and Robert Puentes, "How Lyft and Uber Can Improve Transit Agency Budgets," Brookings Institution report, 8 March 2016. Another report suggests the costs could be even higher, $27 to $38 on average. See Greg Sullivan, "What If 'The Ride' Operated Like the Best Big Paratransit Systems in the US?," Pioneer Institute, 7 April 2015.

32. Kane, Tomer, and Puentes, "How Lyft and Uber Can Improve."

33. Nidhi Subbaraman and Dan Adams, "Uber Boston Announces Partnership with Disability Advocates," *Boston Globe*, 4 February 2016; Nicole Dungca, "MBTA to Subsidize Uber, Lyft Rides for Customers with Disabilities," *Boston Globe*, 16 September 2016.

34. Waylen Miki, "How Will the Sharing Economy Affect Public Transit?," Metro Transit Dispatches, 2 August 2016.

35. Andrew Longeteig, "TriMet Tickets App Now Helps Riders Connect to Other Transportation Options," TriMet News, 5 May 2016.

36. "Uber + DART Means More Complete Transit Trips," DART Daily, 14 April 2015.

37. "MARTA On the Go," itsmarta.com, accessed 4 September 2017.

38. Rahul Kumar, "How the Uber Effect Will Reinvent Public Transit," American City & County Viewpoints, 15 February 2017.

39. Hanjiro Ambrose, Nick Pappas, and Alissa Kendall, "Electric Transit Buses: Transition Costs for California Transit Agencies," Institute of Transportation Studies, University of California, Davis, June 2017.

40. James Ayre, "China 100% Electric Bus Sales Grew to ~115,700 in 2016," CleanTechnica, 3 February 2017.

41. Alissa Walker, "Can Self-Driving Technology Save the Bus?," Curbed, 28 July 2016.

42. Jerome M. Lutin and Alain L. Kornhauser, "Application of Autonomous Driving Technology to Transit: Functional Capabilities for Safety and Capacity," Table 4, Princeton University research paper, Operations Research and Financial Engineering website, 22 July 2013.

43. Danielle Muoio, "This Self-Driving Bus Could Radically Change Public Transportation," Business Insider, 22 August 2016.

44. Bryan Mistele, "Sound Transit's Expansion Will Be Obsolete before It's Built," *Seattle Times*, 9 July 2016.

45. American Public Transportation Association, "APTA's Policy Framework on Integrated Mobility, Transformative Technologies," 3 October 2015. This was adopted by APTA's board of directors.

Chapter 6

1. Nicholas J. Klein and Michael J. Smart, "Car Today, Gone Tomorrow: The Ephemeral Car in Low-Income, Immigrant and Minority Families," *Transportation* 42 (January 2015): 1–16.

2. John Pucher and John L. Renne, "Socioeconomics of Urban Travel: Evidence from the 2001 NHTS," *Transportation Quarterly* 57 (Summer 2003): 49–77.

3. US Census Bureau, 2011–2015 American Community Survey five-year estimates, weighted. Obtained via IPUMS-USA, University of Minnesota, https://usa.ipums.org/usa/.

4. US Department of Transportation, Federal Highway Administration, 2009 National Household Travel Survey, http://nhts.ornl.gov.

5. Ibid.

6. These figures are based on fifteen thousand miles of travel per year. AAA, "Your Driving Costs," AAA NewsRoom, 23 August 2017.

7. US Department of Labor, 2015 Consumer Expenditure Survey, Table 1203.

8. Elaine Murakami and Jennifer Young, "Daily Travel by Persons with Low Income," US Department of Transportation, Federal Highway Administration, 1997; Klein and Smart, "Car Today, Gone Tomorrow."

9. California Department of Transportation, *2010–2012 California Household Travel Survey Final Report* (Austin, TX: Nustats Research Solutions, June 2013).

10. Klein and Smart, "Car Today, Gone Tomorrow."

11. Evelyn Blumenberg and Margy Waller, "The Long Journey to Work: A Federal Transportation Policy for Working Families," Brookings Institution Series on Transportation Reform, July 2003.

12. Bruce Schaller, "Taxi, Sedan, and Limousine Industries and Regulations," Appendix B in Transportation Research Board Special Report 319, *Between Public and Private Mobility: Examining the Rise of Technology-Enabled Transportation Services* (Washington, DC: National Academies Press, 2016).

13. Pauline van den Berg, Theo Arentze, and Harry Timmermans, "Estimating Social Travel Demand of Senior Citizens in the Netherlands," *Journal of Transport Geography* 19 (2011): 323–31.

14. City of Chicago, "Senior Transportation Programs," 2017; San Francisco Municipal Transportation Authority, "Paratransit," 2017.

15. Evelyn Blumenberg et al., "Typecasting Neighborhoods and Travelers: Analyzing the Geography of Travel Behavior among Teens and Young Adults in the U.S.," UCLA Institute of Transportation Studies, October 2015.

16. Elizabeth Kneebone and Emily Garr, "The Suburbanization of Poverty: Trends in Metropolitan America, 2000 to 2008," Brookings Institution Metropolitan Opportunity Series, January 2010.

17. Janet Viveiros and Maya Brenan, "Aging in Every Place: Supportive Service Programs for High and Low Density Communities," Center for Housing Policy, March 2014.

18. Joseph Grengs, "Job Accessibility and the Modal Mismatch in Detroit," *Journal of Transport Geography* 18 (January 2010): 42–54.

19. Taner Osman et al., "Not So Fast: A Study of Traffic Delays, Access, and Economic Activity in the San Francisco Bay Area," University of California Center on Economic Competitiveness in Transportation, 2016.

20. Genevieve Giuliano, His-Hwa Hu, and Kyoung Lee, "The Role of Public Transit in the Mobility of Low Income Households," METRANS Transportation Center, 2001.

21. "Shopping for Used Electric Cars," CARFAX, 22 August 2016.

22. Gil Tal, Michael A. Nicholas, and Thomas S. Turrentine, "First Look at the Plug-in Vehicle Secondary Market," Institute of Transportation Studies, University of California, Davis, Working Paper UCD-ITS-WP-16-02, 2016.

23. David Louie, "Renters Hit Roadblocks to Get Electric Charging Stations," ABC7 News, 31 March 2015.

24. David Diamond, "The Impact of Government Incentives for Hybrid-Electric Vehicles: Evidence from US States," *Energy Policy* 37 (March 2009): 972–83.

25. Tamara L. Sheldon, J. R. DeShazo, and Richard T. Carson, "Designing Policy Incentives for Cleaner Vehicles: Lessons from California's Plug-In Electric Vehicle Rebate Program," *Journal of Environmental Economics and Management* 84 (July 2017): 18–43.

26. Alex Turek and George M. DeShazo, "Overcoming Barriers to Electric Vehicle Charging in Multi-unit Dwellings: A South Bay Case Study," Alternative and Renewable Fuel and Vehicle Technology Program, UCLA Luskin Center for Innovation and South Bay Cities Council of Governments, 2016.

27. J. R. DeShazo, "Improving Incentives for Clean Vehicle Purchases in the United States: Challenges and Opportunities," *Review of Environmental Economics and Policy* 10 (Winter 2016): 149–65.

28. Michael Duncan, "The Cost Saving Potential of Carsharing in a US Context," *Transportation* 38 (March 2011): 363–82; Rosanna Smart et al., "Faster and Cheaper: How Ride-Sourcing Fills a Gap in Low-Income Los Angeles Neighborhoods," BOTEC Analysis Corporation, July 2015.

29. Susan Shaheen, Adam Cohen, and M. Chung, "North American Carsharing: 10-Year Retrospective," *Transportation Research Record* 2110 (2009): 35–44.

30. Steve Hanley, "Low-Income Electric Carsharing Comes to Los Angeles," CleanTechnica, 13 August 2015.

31. Lisa Rayle et al., "App-Based, On-Demand Ride Services: Comparing Taxi and Ridesourcing Trips and User Characteristics in San Francisco," University of California Transportation Center, Research Report UCTC-FR-2014-08, August 2014.

32. Carl Bialik et al., "Uber Is Serving New York's Outer Boroughs More than Taxis Are," FiveThirtyEight, 10 August 2015.

33. FDIC Unbanked/Underbanked Survey Study Group, "2015 FDIC National Survey of Unbanked and Underbanked Households," Federal Deposit Insurance Corporation, Division of Depositor and Consumer Protection, 20 October 2016.

34. Aaron Smith et al., "U.S. Smartphone Use in 2015," Pew Research Center, April 2015.

35. NYC Department of Consumer Affairs, Office of Financial Empowerment, "Immigrant Financial Services Study," November 2013.

36. Chicago Transit Authority, "Ventra," 2016.

37. Brent Begin, "Car Sharing Rare in San Francisco's Lower-Income Areas," *San Francisco Examiner*, 28 July 2011.

38. Elliot Fishman, Simon Washington, and Narelle Haworth, "Bike Share: A Synthesis of the Literature," *Transport Reviews* 33 (2013): 148–65.

39. Yanbo Ge et al., "Racial and Gender Discrimination in Transportation Network Companies," National Bureau of Economic Research, NBER Working Paper No. 22776, October 2016.

40. Kevin Hickman, "Disabled Cyclists in England: Imagery in Policy and Design," *Proceedings of the Institution of Civil Engineers—Urban Design and Planning* 169 (June 2016): 129–37.

41. "Westminster and Zagster Introduce Inclusive Bike Share Program," Zagster, 30 June 2016.

42. Alyson Kay, "College Park Is Leading Other Cities with Its Disability-Friendly Bike-Share Program," *Diamondback*, 13 September 2016.

43. Daniel Ward, "Transportation Network Companies and Accessibility," issuu.com, 22 August 2016.

44. Michael Theis, "RideScout Trip-Planning App to Soon Be Obsolete," *Austin Business Journal*, 29 July 2016.

45. Erick Guerra, "Planning for Cars That Drive Themselves: Metropolitan Planning Organizations, Regional Transportation Plans, and Autonomous Vehicles," *Journal of Planning Education and Research* 36 (June 2016): 210–24.

Chapter 7

1. Levi Tillemann, "Crossing the Divide: The Long Arc of Corporate Efficiency," valencestrategic.com, February 2017.

2. Joseph Alois Schumpeter, *Capitalism, Socialism, and Democracy*, 3rd ed. (New York: Harper, 1950).

3. Mike Isaac, "Uber's C.E.O. Plays with Fire," *New York Times*, 23 April 2017.

4. Robert Ferris, "Tesla Passes General Motors to Become the Most Valuable U.S. Automaker," CNBC, 10 April 2017. At that time, Uber was valued at close to $70 billion, but it had not yet gone public.

5. Robin Chase, *Peers Inc.: How People and Platforms Are Inventing the Collaborative Economy and Reinventing Capitalism* (New York: Public Affairs, 2015).

6. James P. Womack, Daniel T. Jones, and Daniel Roos, *The Machine That Changed the World* (New York: Free Press, 1990).

7. Ibid.; Jeffrey K. Liker, *The Toyota Way: 14 Management Principles from the World's Greatest Manufacturer* (New York: McGraw-Hill, 2004).

8. Todd Lassa, "Toyota, Chrysler Have North America's Most Efficient Plants," *Motor Trend*, 5 June 2008.

9. James R. Healy, "Average New Car Price Zips 2.6% to $33,560," *USA Today*, 4 May 2015.

10. Paul Barter, "'Cars Are Parked 95% of the Time.' Let's Check!," Reinventing Parking, 22 February 2013.

11. Adella Santos et al., *Summary of Travel Trends: 2009 National Household Travel Survey* (Washington, DC: Federal Highway Administration, 2011), Table 16.

12. This concept originated with my company, Valence Strategic, and should be partially credited to my business partner Colin McCormick.

13. See the Airlines for America website, http://airlines.org/.

14. Tillemann, "Crossing the Divide."

15. James Archsmith, Alissa Kendall, and David Rapson, "From Cradle to Junkyard: Assessing the Life Cycle Greenhouse Gas Benefits of Electric Vehicles," *Research in Transportation Economics* 52 (2015): 72–90; M. A. Tamayao et al., "Regional Variability and Uncertainty of Electric Vehicle Life Cycle CO_2 Emissions across the United States," *Environmental Science and Technology* 49 (2015): 8844–55.

16. Neal E. Boudette, "Automakers Prepare for an America That's over the Whole Car Thing," *New York Times*, 22 December 2016.

17. John Cook, "Ford CEO Mark Fields Sees 'Massive Opportunity' to Push Car Maker into $5.4T Transportation Services Market," GeekWire, 5 January 2016.

18. Michael Martinez, "Ford Motor Wants to Change the World—Again," *Automotive News*, 10 January 2017.

19. Ford Motor Company, *Sustainability Report 2015/16*, accessed 4 September 2017, http://www.sustainability.ford.com.

20. Nick Bunkley, "Ford's Mobility Push Driven by Profit Motive," *Automotive News*, 19 September 2016.

21. Maven website, accessed 4 September 2017, https://www.mavendrive.com/#!/.

22. Peter Kosak, "GM on Fast Track to Redefine Mobility with Lyft, Maven & Self-Driving Cars," General Motors Green, 24 March 2016; John Rosevear, "Is General Motors' New Deal with Uber a Stab in Lyft's Back?," Business Insider, 4 November 2016.

23. Nick Bunkley, "GM Puts Bolt at Center of Its Long-Term Vision," *Automotive News*, 19 December 2016.

24. John Branding, personal communication, 10 April 2017.

25. "Uber's Revenue Hits $6.5 Billion in 2016, Still Has Large Loss," Reuters, 14 April 2017.

26. Dana Hull, "Elon Musk Says Tesla Car-Share Network Is 'the People vs. Uber,'" Bloomberg Technology, 26 October 2016.

27. Boudette, "Automakers Prepare."

28. Andrew Salzberg, personal communication, 13 April 2017.

29. Elisabeth Behrmann, "Daimler Looks to Startups, Small Suppliers for Innovative Ideas," *Automotive News Europe*, 1 March 2017.

30. Eric Auchard and Laurence Frost, "BlaBlaCar Unveils Opel Leasing Deal in Boost for Ride-Sharing," *Automotive News Europe*, 6 April 2017.

31. Hans Greimel, "Toyota's Growth Model Turns on Services," *Automotive News*, 7 November 2016.

Chapter 8

1. World Health Organization, WHO Global Urban Ambient Air Pollution Database, 2016 update.

2. J. Lelieveld et al., "The Contribution of Outdoor Air Pollution Sources to Premature Mortality on a Global Scale," *Nature* 525 (17 September 2015): 367–71.

3. GBD MAPS Working Group, "Burden of Disease Attributable to Coal-Burning and Other Air Pollution Sources in China," Health Effects Institute, Special Report 20, August 2016.

4. "United States Vehicle Ownership Data, Automobile Statistics and Trends," Hedges & Company, https://hedgescompany.com/automotive-market-research-statistics/auto-mailing-lists-and-marketing, accessed 4 September 2017.

5. Chinese sales of new energy vehicles in 2016 totaled 507,000, consisting of 409,000 all-electric vehicles and 98,000 plug-in hybrid vehicles (many of these trucks and buses), according to Liu Wanxiang, "中汽协: 2016年新能源汽车产销量均超50万辆,同比增速约50%" [China Auto Association: 2016 New Energy Vehicle Production and Sales Were over 500,000, an Increase of About 50%], D1EV, accessed 12 January 2017.

6. Brian Solomon, "China Legalizes Ridesharing, Opening the Door for Uber, Didi," *Forbes*, 28 July 2016.

7. Li Fusheng, "Road Map Outlined for New Energy Industry," *China Daily*, 31 October 2016.

8. Paul Gao et al., "Finding the Fast Lane: Emerging Trends in China's Auto Market," McKinsey & Company survey, April 2016.

9. François-Joseph Van Audenhove et al., "The Future of Urban Mobility 2.0," Arthur D Little Future Lab, January 2014.

10. Mark Kane, "China Revises Policy for Foreign Battery Manufacturers—Floodgates Now Open," InsideEVs, 20 July 2016.

11. Jose Pontes, "China Electric Car Sales Demolish US and European Electric Car Sales," CleanTechnica, 25 January 2017.

12. Eric Loveday, "Wanxiang Gets Final Approval to Buy A123 Systems," PluginCars, 21 January 2013.

13. Yunshi Wang et al., "Overview of China's New Energy Vehicle (NEV) Regulation and Implications for California Market," Institute of Transportation Studies, University of California, Davis, November 2017.

14. Statista, "Number of Employees at Baidu from 2009 to 2016," accessed 4 September 2017.

15. Danny Shapiro, "Baidu's Project Apollo Takes Flight, Bringing Autonomous Cars Closer to Reality," Nvidia, 5 July 2017.

16. Michael Dunne, "Baidu's Bet on Beijing's Autonomous Car Drive," *Nikkei Asian Review* 11 (May 2017).

17. Katie Fehrenbacher, "Padmasree Warrior on the Future of Transportation and Tesla," Fortune Tech, 15 October 2016.

18. Fred Lambert, "NextEV Will Build a $500 Million Factory to Manufacture Electric Vehicle Motors in China," Electrek, 5 May 2016.

19. Tycho De Feijter, "LeEco to Build $3B Electric Car Factory in China," *Forbes*, 11 August 2016; Bryan Logan, "'We Are in a Precarious Situation': The Electric-Car Startup Faraday Future Is Scrapping Its Big Nevada Factory as Its Cash Crisis Deepens," Business Insider, 10 July 2017.

20. Kristen Hall-Geisler, "A Ride in the Lucid Motors Alpha Prototype," TechCrunch, 4 January 2017.

21. Aaron Mamiit, "China's Baidu Unveils New, All-Electric Self-Driving Car: Testing Begins for Modified Chery EQ," Tech Times, 28 August 2016.

22. Zheng Wan, Daniel Sperling, and Yunshi Wang, "China's Electric Car Frustrations," *Transportation Research Part D* 34 (2015): 116–21.

23. Steve Hanley, "China EV Sales, Subsidies, and Targets," EVObsession, 2 May 2016.
24. Yunshi Wang et al., "China's Electric Car Surge," *Energy Policy* 102 (March 2017): 486–90.
25. "China Fines Five Auto Makers for Electric-Vehicle Subsidy Fraud," *Wall Street Journal*, 8 September 2016.
26. Fusheng, "Road Map Outlined."
27. "Uber Losing $1 Billion a Year to Compete in China," Reuters, 18 February 2016.
28. Alejandro Zamorano, "Digital Ride Hailing: The Global Landscape," Intelligent Mobility Research Note, Bloomberg New Energy Finance, 22 August 2017.
29. Paul Traynor and Didi Tang, "Chinese Using Carpooling Apps to Get Ride Home for Holidays," *San Diego Union Tribune*, 4 February 2016.
30. "25-Year-Old's $500 Million Startup Fuels China Bike-Share Battle," Bloomberg Technology, 30 October 2016.
31. George Chen, "When 60,000 Taxis Are Not Enough for Shanghai," *South China Morning Post*, 26 July 2015.
32. "Shanghai Metro," Wikipedia, last modified 31 July 2017.
33. Zhang Yi, "Traffic Gets Heavier in Third of Cities: Jinan Ranks Worst," *China Daily*, 12 January 2017.
34. Meng Jing and Zhu Lixin, "Baidu Gets Green Light to Set Up Test Centre for Driverless Vehicles," Asia News Network, 17 May 2016.
35. "Chinese City Wuhu Embraces Driverless Vehicles," BBC News, 16 May 2016.

About the Contributors

Anne Brown is a researcher at the Institute of Transportation Studies and a PhD student in urban planning at the Luskin School of Public Affairs at UCLA. Her research focuses on travel behavior, transportation finance, and equity.

Robin Chase is a transportation entrepreneur. She cofounded Zipcar, the world's largest carsharing company, and Veniam, a vehicle communications company. She is the author of *Peers Inc.: How People and Platforms Are Inventing the Collaborative Economy and Reinventing Capitalism* (Public Affairs, 2015) and lectures widely on innovation, entrepreneurship, technology, transportation, cities, and climate change. In 2009, she was honored by *Time* magazine as one of the year's 100 Most Influential People. She serves on the board of directors of Veniam and Tucows and on the advisory board of the World Resources Institute.

Michael J. Dunne writes about cars, Asia, and the future of mobility. A Detroit native and an entrepreneur with twenty-five years in Asia, Dunne founded Dunne Automotive Ltd. in 2015 to advise investors on the most promising areas for growth and profits. Dunne's commentaries have been published in the *Wall Street Journal Asia*, *Forbes*, the *New York Times* Global Edition, and *Automotive News*. Dunne is also the author of *American Wheels, Chinese Roads* (Wiley, 2011).

Susan Pike, PhD, is a postdoctoral researcher at the UC Davis Institute of Transportation Studies. Her areas of interest include travel behavior, sustainable transportation, environmental policy, and future mobility and emerging transportation services.

Steven E. Polzin, PhD, is a professor of civil and transportation engineering and director of mobility policy research at the Center for Urban Transportation Research at the University of South Florida. His research focuses on public transportation, public policy analysis, transportation planning, systems evaluation, planning process design, and mobility analysis. He is on the editorial board of the *Journal of Public Transportation* and serves on several Transportation Research Board and APTA committees.

Susan Shaheen, PhD, is an adjunct professor of civil and environmental engineering at UC Berkeley with an interest in socially and environmentally beneficial technology applications. She codirects the Transportation Sustainability Research Center and directs the Innovative Mobility Research group at UC Berkeley. She received the 2018 Roy Crum Award from the Transportation Research Board for her leadership and accomplishments in advancing new fields of mobility research.

Daniel Sperling, PhD, is founding director of the UC Davis Institute of Transportation Studies and codirector of the National Center for Sustainable Transportation at UC Davis. He is a distinguished professor of civil and environmental engineering and environmental science and policy at UC Davis and is a member of the influential California Air Resources Board. He has authored or coauthored more than 250 technical papers and reports on transportation and energy, along with 12 books, among them *Two Billion Cars: Driving toward Sustainability* (with Deborah Gordon; Oxford University Press, 2009) and *Future Drive: Electric Vehicles and Sustainable Transportation* (Island Press,

1995). Professor Sperling contributed to the reports of the Intergovernmental Panel on Climate Change, which won the Nobel Peace Prize in 2007, and he was awarded the Blue Planet Prize in 2013 for bringing science to policy.

Brian D. Taylor, PhD, is a professor of urban planning and director of the Institute of Transportation Studies and the Lewis Center for Regional Policy Studies in the Luskin School of Public Affairs at UCLA. His research centers on travel behavior, transportation finance, social equity, traffic congestion, and shared mobility, and he has authored or coauthored more than a hundred journal articles, reports, and monographs on these topics. He was named one of the top ten academic thought leaders in transportation by the Council of University Transportation Centers and the Eno Center for Transportation in 2016.

Levi Tillemann, PhD, is managing partner at Valence Strategic and a fellow at New America. He is the author of *The Great Race: The Global Quest for the Car of the Future* (Simon & Schuster, 2015), a book on the intersection of policy and innovation in the global auto industry. Previously he served as special advisor for policy and international affairs at the US Department of Energy and was CEO of IRIS Engines, a company he founded to develop a smaller, more efficient, and more powerful combustion engine (based on a design for which he holds multiple patents). His writing has appeared in the *Washington Post*, the *Wall Street Journal*, *Fortune*, and the *New Yorker*, and he is a regular guest on radio and television programs, including NPR's *Diane Rehm Show* and *Marketplace*.

Ellen van der Meer was a visiting graduate scholar at the UC Davis Institute of Transportation Studies in late 2016. She earned an MSc in innovation sciences from Utrecht University in the Netherlands and is focusing her research on the dynamics of emerging technologies and innovation.

Index

Figures/photos/illustrations are indicated by an "f" and tables by a "t."